DARK ENERGY

Alex J Morrey BSc.

ISBN: 1452829527
ISBN-13: 9781452829524
Library of Congress Control Number: 2010905911

TABLE OF CONTENTS

Dark Energy Foreword

Science and mysticism could not be further apart in the modern way of thinking. They are mutually exclusive and completely opposite from each other. Science depends on an objective and measurable view of the universe. Mysticism results from intuition relating to various myths, legends, and visions. The two could not be further apart and opposite from each other in most people's minds. Alex J Morrey developed in previous writings a new model to unite science and mysticism into one unified theory. In "Dark Energy," he further develops his model to the point where it approaches the long sought after, "Theory of Everything."

When I was in the fifth grade, I asked my dad how gravity worked. He introduced me to the standard equation normally used to define Newton's law. After some effort, he caused my young mind to understand the math. But he was not able to answer the real question that I was trying to understand. We can see that gravity follows the law defined by the math, but why is that? Where does gravity come from? Why did it decide to follow that equation, and not another equation? All he could tell me is that was the way God designed it. My dad was a great mathematician and scientist to my eleven year old mind. He could cause me to understand anything. Yet when I drilled my questions down to the root, he could offer only a religious answer, even though he was not a religious man. Like any other scientist, he could show me the equations that describe how the physical world behaves, but he could not tell me why it behaved that way.

Alex's model answers those basic questions. My dad was correct. God did design it that way. But his model of God is far different from the orthodox view. I'm a scientist by education and an engineer by profession. I love both with a deep and burning love. I'm also a Christian mystic, which I find to be completely compatible with the science and engineering. My chosen mystical tradition is based upon myths, legends, and parables—none of which is to be taken literally, but all of which point toward a very deep understanding of the universe. This is the spiritual language by which mystics communicate with each other because normal human language is completely inadequate for the task. Every wisdom tradition develops its own spiritual language, which initiates must learn before they can learn the deeper truths. Once the spiritual languages of all the traditions are understood, it can clearly be seen that they all teach the same thing with only very minor variations. They are equal to each other in every respect.

Alex's model is essentially a new spiritual language teaching the same ancient wisdom, and unifying it with modern science. He has developed his model not on the myths and legends of old, but rather on the modern language and understanding of science. He accepts all that modern science teaches, but he goes deeper and further than any of the standard objective models are able to go. As my early experience with gravity and my dad showed, science does an excellent job of explaining how things work; but if you drill your questions down to the basic roots, you will always be left with a question that conventional science cannot answer: "Why does it work that way?" Alex answers that question.

Paul L Kruse, Merritt Island, Florida, USA. E-mail plkruse@cfl.rr.com 28th June 2010.

Preface

You may wonder how this line of thinking came about. Let me explain. As a university student in 1959, I was given a book to read by a friend. The book was entitled *The Power of the Mind* and was written by Dr. Rolf Alexander, MD, in 1956 The book opens with the statement that our mind is our world:

> Your world is your mind. Without a mind to behold it, the world would not exist. The colouring of the autumn woodlands, the witchery of moonlight over a woodland lake, the blazing desert sunset, the face of a loved one, in fact, every experience of your life is a construction of your mind, an interpretation of stimuli arriving from without. Heat and cold, light and darkness, form, velocity, colour, sound, taste, feeling and emotion cause responses in the human mind. Yet the response of each mind to these stimuli is as individualistic as are a man's fingerprints.

This thesis remains the central point of his discussion, which goes on to explore a self-training auto-hypnosis readers can achieve, during which they can implant suggestions into their own belief structure in order to modify their perceptions of the world in which they live, with dramatic consequences. This means, for example, that someone who has long suffered from debilitating allergies can, in part, reprogram his own mind so that this idea becomes incompatible with his new belief structure and subsequently ceases to be real, leaving him allergy-free. The applications of such reprogramming are literally endless, but bear in mind that this method of healing can only succeed if the affliction is a hallucination lying beyond the influence of the coercive interaction, as we shall continue to discuss throughout this book..

The main purpose of the hypnotic training, however, is to get the student to become fully aware and cognizant of being in the trance or dream state; he will then be able to realise that his everyday world is also a dream state, and will seek to live more fervently in Reality.

"At last somebody is talking sense," was my first thought on reading the book. It was like a breath of fresh air going through my mind. In *The Power of the Mind,* Dr. Ralph Alexander stresses the point that a very great deal of what we believe to be unchangeable, immoveable truth is actually in our own minds. We effectively create the world in which we live and each one of us inhabits a world ever so slightly different from those in which all other people find themselves. After reading the book, I put my thoughts on it aside for the time being and got on with life, work in the form of study and a myriad of trivial distractions. However, about 18 months later, when I was walking along Ventnor Beach at Wheelers Bay on the Isle of Wight, a great part of my mind was suddenly reconfigured, leaving me utterly astounded and fully aware of the truth of Dr. Alexander's message. I had thought that he made an excellent point while I was reading the book, but now I knew that his thesis was largely correct: *each of us creates, to some extent at least, the world in which we live.* According to Dr. Alexander, a realization of this nature should happen to anybody whose mind is able to grasp what he is implying. This is a profoundly liberating experience, as one becomes aware that one's destiny, future and present are all in one's control and not at the mercy of the "slings and arrows of outrageous fortune."

I was so excited by this realization that I wrote on and off in my notebook for three days and could hardly sleep. Knowing that I had much more power to influence my own future and present state than I had ever imagined, I felt almost as though my hand was writing all by itself. It was an amazing sensation, as if the action had nothing to do with me. Now, the writing was not very good and nor was the spelling, but it was quite an experience! I still have that notebook.

"In those days" I was studying physics at university and was very interested in the nature of matter, that stuff out of which all of the objects in the world around us are made. As we have seen, this was an interest that would later develop in several unpredicted directions. Back then, I glanced at a book about Hinduism in the university library which had been written by a Dr. Albert Schweitzer. I found the topic rather odd because he was a Christian missionary who worked in Africa somewhere. On the first page, Schweitzer posed the question, "What is Hinduism?" His answer was, "Hinduism is the search for truth."

Now, I had thought that *I* was searching for the truth by studying physics, which seemed billions of light years away from Hinduism, so I was most intrigued by this proclamation. Religion the search for truth? How could that be when science, as far as I was then concerned, was the business of understanding the true nature of things? What possible insight could religion—*any* religion—have into the matter? Like so many people of a primarily scientific bent, the study of religion and the quest for knowledge seemed to be two quite separate courses, which could never intersect. I set the matter aside for the time being. Not long thereafter, I read yet another book, written by Dr. Rhine of Duke University about his experiments with extra-sensory perception and other "gifts" generally considered to belong firmly in the realm of the paranormal. I knew almost nothing about such things at that time but reading about them left me rather puzzled. With a background in physics, I just could not comprehend how the "Laws of Nature" could be bent, or even broken, by individuals possessing such "gifts."

Things have moved on quite a long way since those days.

Alex J Morrey 26th May 2010

Introduction

I don't believe that the equations that predict dark energy, dark matter and dark flow are incorrect. Rather, the current model is inadequate and needs upgrading. That is exactly what is intended by this work. It's bound to be mind-bending at first sight because it's necessary for us to be able to think outside the box, the paradigm, into which we have been "educated."

Current scientific understanding tends to assume the objective nature of what is perceived to be the universe, the world in which we are living. This means that the universe is assumed to have come into existence at some time in the past, referred to as "the beginning," the so-called Big Bang or something. Others who may or may not be scientists believe that the universe and everything within it was created miraculously by God. There are still others who prefer a so-called Quantum Mechanics explanation. In this work an interesting forth option is being put forward.

There is no argument about the reality of the world in which we are living; it is real. The question is: Why is it real, or what makes it real? Such a question seems irrelevant if the universe is assumed to be objective fact.

As scientists when we think about the universe as a consequence of objective fact we seem to be unable to obtain satisfactory answers to some very fundamental questions on which everything else depends:

1) How did the universal come to be?

2) What is it made of? Matter? But what *is* matter?

3) How was life bestowed on inanimate matter before the process of evolution could begin anywhere in the universe?

4) What is the so-called Dark Energy - the dark matter and dark flow that fill the universe?

Nor will we be able to properly understand issues such as:

1) Darwin's evolution by natural selection.

2) What is the meaning and purpose of an individual human life beyond the mundane?

3) Is there life after death?

4) Is it possible to predict the future Nostradamus style, or to travel through time?

5) Is telepathy possible?

6) Are so-called UFOs and extra-terrestrial life-forms real?

7) Is it possible to travel faster than the speed of light?

8) What is the so-called Multiverse, dark matter and dark flow?

Then there are questions that we think that we know the answers to, such as that of Gravity. Even this needs a bigger picture:

The gravitational field of the earth and "walking" on water?

The electrostatic field and the nature of matter.

The real cause of global warming and climate change:

Is there really a God out there?

A very different approach is necessary.

If we consider that the universe is *perceived* to be real because we have collectively come to "unconsciously believe" in its reality, this does provide the answers to the fundamental questions, although they're not quite the kind of answers we are expecting. In order to do this we need to upgrade our current understanding to include the concept of Reality and we need to understand with our unconscious and our conscious minds working together as one integrated mind. To "unconsciously believe" is quite different than to "consciously believe" as a thought belief. It's a matter of thinking outside the box into which we are indoctrinated by the so-called education system. The more highly educated we become, the more difficult it is to do this.

We shall see as this work progresses that Reality is outside space/time in the "gap" between yesterday, tomorrow and elsewhere. The reality of the present, the past and the future are all being "engendered" into existence from microsecond to microsecond by humanity itself by means of a subliminal "coercive interaction" between all the unconscious minds of humanity in conjunction with Reality.

The reality of our world in its current state is a conflicting "logical" structure that I refer to as the universal belief (UB) that has been 'created' or, more precisely, 'engendered' by a subliminal perceptual consensus between all the minds of humanity. The conflicting structure of reality shows itself as all the woes of humanity. A belief such as this is something like software running within a computer. However, a computer has no consciousness with which to perceive its "belief", unlike a human mind with the "software" of the UB running within it or, more precisely, an instance of the software of the UB running within it. This is perceived by the innate consciousness, within a human brain, as real and solid and is in some ways similar to an individual sleeping dream-world within an individual human mind. In this case, however, it is not a dream as such but a dream-like world within the total collective mind of humanity in which the brain together with its body perceives itself to be living, the UB.

The so-called Laws of Nature will operate just as well within the universal belief as they do within the objective universe; genetics, for example, will work just as well within the universal belief as they do in the assumed objective universe. In fact, we would not know the difference. Facts will be the same facts in both interpretations but, in the first case, facts will be facts of unconscious belief instead of assumed objective facts. For example, so-called material objects in the objective universe are perceived to be real objects, but within the UB the same objects are actually made of the same "stuff" of which objects within a human mind are also made. Even so, they are also perceived as real objects that feel solid to the tactile senses, having inertia and mass. They are also considered to be factual objects but are, in this case, a fact of unconscious belief.

In order to understand how so-called "imaginary objects" within a human mind can exhibit inertia and mass and feel solid requires a different understanding of the so-called electrostatic field and laws of motion. Another such fact is the so-called law of gravity. Yet another is the speed of light, being 299,792 kps. Facts are based on axioms; self-evident truths that are assumed. We can't ask why the speed of light is 299,792 kps. In a universe that is considered to be objectively real, this has to be considered a self-evident truth that we have no alternative but to accept as fact. Within the UB, however, that question can be asked and answers found. There is a very subtle difference between the workings of the objective universe and the workings of the UB, but the implications are far-reaching. There are also staggering religious implications.

There is no way of determining which one of the two options we are actually living in by studying natural events in space and time alone. The same mathematics work equally well in both the UB and the objective universe. However, when we come to understand that the UB can be enhanced into a lower conflicting state that allows the so-called Laws of Nature to be temporally "upgraded," under certain circumstances, before they are perceived as reality, our understanding is upgraded, too. The law of gravity, for example, can be enhanced so that, in due course, it is perceived as a zero G effect, allowing certain people living close to Reality to walk on water or levitate.

This approach allows science to understand and work not only with natural events, as it currently does, but also with so-called supernatural events. Of course, within this new way of thinking, such events will no longer be considered supernatural. The so-called theory of the universal belief gives science a very much bigger "base" in which to work. We need to develop a deep understanding of the nature of the UB that is perceived as reality and how it came to be, and its ultimate potential development into Reality.

Matter and Consciousness

What is matter? We think of the objects in our everyday world as being composed of an array of atoms or even, nowadays, so-called "strings," in a fairly stable state of dynamic equilibrium within the envelope of a material object. If required please view articles on Wikipedia, the online encyclopaedia, for more information on String Theory. This means that all objects are composed mostly of space and time, the space and time between the atoms. If we were to stop at this point, we could say that matter is composed mostly of space and time. However, we would be left with "bits" of matter in the form of atoms floating around within this space and time and we would not be answering the question:"What is matter"? We must pursue this question further.

Within this more compacted space and time, we think of atoms as much smaller objects behaving in much the same way as objects in the normal space and time of our everyday world (this is not strictly accurate but it will serve our purposes for the time being.). In turn, we think of atoms as consisting of still smaller objects in the form of a central nucleus surrounded by an array of electrons, also in a fairly stable state of dynamic equilibrium around the central nucleus—more stable, in fact. This means that atoms are composed mostly of space and time more densely packed than that surrounding them.

We think of a nucleus as consisting of an array of protons and neutrons in a highly stable state of dynamic equilibrium within a still more densely compacted space and time. We proceed to think of protons and neutrons, etc., which make up the nucleus of an atom, as consisting of even smaller material objects called quarks existing within an even more densely compacted space and time, within the envelopes of the protons and neutrons and in an extremely stable state of dynamic equilibrium. Electrons are "known" to have dimensions. This means that they also consist of even smaller material objects within a more highly compacted space and time, within the envelope of the electron. Continuing our train of thought to its logical conclusion, we will eventually arrive at the point where matter consists of an array of geometrical points in a state of dynamic equilibrium; a geometrical point which, although it has no dimensions and no mass, can be thought of as the ultimate smallest "particle." We could try to suggest, in keeping with the latest quantum passion of science, that it is a very small point of energy and start talking about "strings," but energy and mass are equivalent so it would not be a geometrical point.

Introducing the Universal Belief

All of this implies that matter is composed entirely of space and time, which in turn means that the entire universe is composed of space and time, making it a logical system rather like software running in a vast super-computer, the logic of which could, in theory, be expressed totally by some form of mathematics. In the context of this work, even this is not the total picture. We still need to include those elusive entities we experience as "consciousness" and "will." With the addition of these important ingredients, we can view the universe as a sort of mind: an entity incorporating everything, including each and every one of us, and functioning totally below the level of awareness. In fact, the perception of the real world in which we live, the universe, stems from the belief structure of our own individual minds/brains. This is what is referred to when the term "Universal Belief" (UB) is used throughout this work. The Universal Belief is not conscious of the universe, it *is* the universe, and it functions on a purely unconscious level. When the UB or, more precisely, an instance of it, is perceived by the innate consciousness within an individual human brain it is perceived to be the world in which that brain together with its body or real self is living. It takes the total collective unconscious minds of humanity to contain the entire UB.

An unconscious belief is like software running within a computer but a computer has no innate consciousness with which to perceive its own belief. The UB "running" below the level of awareness within a human brain is consciously perceived from instant to instant by the innate consciousness within a human brain as the real world in which the "brain/body" – the "real self" – is living, although it takes the whole collective mind of humanity to contain the entire UB as mentioned above. This implies that "real material objects" in the so-called real world are actually made of the "stuff" of which "imaginary objects" are made within a human imagination. This being so, we need to ask how such "imaginary objects" can posses inertia and mass and feel solid to the tactile senses. After all, when we are in a sleeping or hypnotically induced dream, objects in that dream are solid to the sense of touch and yet when we wake up from the dream we realise that it was all in our own head. However, reality is not a dream because the whole of humanity is involved in what amounts to dream-like world. Nonetheless, there is a way to wake up from this as well.

The so-called laws of motion, hence the property of matter called inertia, are built into the UB just as we think that they are within the "objective universe". But in the former case the laws of motion, being a belief, can be "enhanced" which means "changed" under certain circumstances. To understand how imaginary objects can possess tactile sensibility requires a different understanding of the so-called electrostatic field.

What we do not realise and can never be aware of directly is that there is a continuous unconscious interaction between all the minds of humanity and what I refer to as Reality. This interaction creates the expanding reality of the world in which we live. From the cosmological point of view, it is the so-called dark energy that fills the universe.

This book examines the answers to the ultimate questions and challenges us to think beyond accepted ways of conscious belief.

The Electrostatic Field

To fully understand the nature of the electrostatic field, it is necessary to understand what we perceive to be the universe in a very different way than we currently do. Instead of assuming the universe to be consequence of concrete, objective fact, that has been in existence from the beginning, whatever that means, a universe composed of atoms, photons and sub-atomic particles within a fabric of space and time, it is considered here to be a consequence of collective unconscious belief, referred to as the universal belief (UB). Atoms, photons and subatomic particles, etc. are real because we have collectively and unconsciously come to believe in their reality.

The universal belief (UB), which is like software running in a computer, has actually been "created" by humanity itself in conjunction with what I refer to as Reality, as shown in my various articles that elucidate the so-called Theory of the Universal Belief. It is "created" or, more precisely, "engendered" by a coercive interaction between all the unconscious minds of humanity in the form of a perceptual consensus. This coercive interaction is the so-called dark energy that fills what we perceive to be the universe. This doesn't make any sense until we can think in terms of a universe of unconscious belief as opposed to a universe of objective fact.

Ordinarily, there is no way of knowing, by observing events in space and time, which of the two options we are actually living in, because the so-called laws of nature work equally in both situations. However, within the universal belief, the laws of nature and even history itself can, under certain conditions, be changed by "enhancement." We also need to take into account those elusive entities experienced as consciousness and will. The so-called universal (UB) belief is like software running in a computer but a computer has no innate consciousness with which to perceive its belief. A human mind, however, does.

The UB or, more precisely, an instance of running within a human brain is perceived by the innate consciousness, within the brain, as real and solid and is in some ways similar to an individual sleeping dream-world within an individual human mind. In this case, however, it is not a dream as such but a dream-like world within the total collective mind of humanity in which the brain, together with its body, perceives itself to be living.

In this work, the "material objects" in the objective universe that are perceived as real objects are also perceived as real objects within the universal belief, even if they are made of the same stuff as imaginary objects within a human imagination. This being so, we need to ask how such imaginary objects can feel so completely solid to the tactile sense, as well as possessing inertia and mass. To answer this question, we need a deeper understanding of what is perceived to be the electrostatic field.

The best way to start thinking about the "electrostatic field" is to consider a hypothetical air traffic control system. Each aircraft has a pilot and an auto pilot, as usual; a navigation system controlled partly from the air traffic control system on the ground and also from within the aircraft itself. In addition, there is a collision avoidance system controlled entirely by the air traffic control system. When two aircraft are on an intersecting collision course, the collision avoidance system takes over the flight controls within each aircraft, diverting each from its originally planned course. This implies that it would be impossible for two aircraft to collide, regardless of what the auto pilot or the pilot tried to do. If the pilot was unaware of the existence of the collision avoidance system, he could easily be led to believe that each aircraft was surrounded by a repulsive field referred to as an electrostatic field. So it is for every "imaginary object" within the universal belief (UB), right down to the atom and sub-atomic particles. The anti-collision software "running" in an unconscious belief such as the UB is like software running within a computer network. The so-called electrostatic laws will be installed into the universal belief as if there were an electrostatic field. The software within the UB would get each object to behave as if it was in an electrostatic field without there being a electrostatic field as such. However, computers have no conscious awareness with which to perceive that belief, unlike the consciousness within each individual human brain within the collective unconscious mind of humanity. Each individual brain running an integrated instance of the UB within itself perceives that instance as the world in which it is living, although it takes the total collective mind of humanity to contain the entire UB, much like a computer network that contains the entire database used by each computer within the network.

Each object within the UB moves and is positioned by the software running within the UB according to the so-called "laws of nature" that have been programmed into the software of the UB. One such rule is that no two "imaginary objects" can occupy the same space/time which means that if two are on a collision course, both will be deflected away from their original trajectories according to the rule or law that has become "programmed" into the UB in the form of the laws of "electrostatic repulsion", but without implying the existence of an electrostatic field.

Our own "physical" bodies are also part of the UB. When the fingertips belonging to a human hand within the UB come into contact with any other object, although the fingertips and the object within the UB are both insubstantial in essence, the software of the UB moves and positions the fingertips and the touched object according to certain rules we refer to as the "laws of nature". When this action occurs within an instance of the UB "running" within an individual human brain, it is perceived in due course by the consciousness innate within the brain as an action in the "real" world in which that particular brain within a particular body (real self) is living. This means that the so-called laws of the electrostatic field are a consequence of an unconscious belief in which we have all come to collectively believe by subliminal perceptual consensus, as opposed to a universe of concrete objective fact. Later we shall come to understand that such laws can be temporarily changed by enhancement under certain circumstances, as can the so-called law of gravity.

The usual way we think of tactile sensation and the solid nature of "material" objects is based on our belief that all objects are composed of atoms that are surrounded by negatively charged electrons, which means that all objects that are made of atoms are surrounded by negatively charged electrons and repel each other according to the electrostatic law of repulsion which, in this case, produces a very short range repulsive force when atoms come into close contact. This is because the outer shell of negatively charged electrons around each atom will become the dominant force that will repel atoms that come into close "contact". This is because the neutralising effect of the positively charged proton within the nucleus will still be a radius away from the point of close "contact" between two colliding atoms. This is the case when human fingertips come into contact with (touch) a "material" object. This is the electrostatic field at the micro level, as we think currently.

As I point out in Dark Energy as well as *Where Angels Fear to Tread*, atoms, photons and sub-atomic particles do not actually exist in or very close to the ultimate reality referred to as Reality because they too are a consequence of unconscious collective belief and are also part of the reality that is the UB. There is no electrostatic field as such. Much the same applies to inertia and mass. They too are functions programmed into the UB, as opposed to objective laws of nature. When an object within the UB is moved, it does so according to the so-called laws of motion programmed into the UB. The so-called past is also an integral part of the UB and can also be enhanced by minds living closer to Reality than other minds around them. This is how we can imagine that reality, together with atoms and photons, came to be real within the universal belief. There is only here and now in reality as well as in Reality. Space and time are also a functional part of the UB, as opposed to the space and time of objective fact.

The so-called supernatural event of a real person walking through a real wall can be explained by the enhancement of the electrostatic field, which makes it possible for two "real objects" to occupy the same space/time event. These ideas do not imply that the science of the 21st century is incorrect but rather gives science a bigger picture in which to work by allowing it to understand not only so-called natural events, but also what used to be called "supernatural". Now we can understand every event as natural.

The Nature Of Reality

The world in which we live, the universe in the greater sense, is considered to be a consequence of collective belief; an unconscious belief much like software running within in a computer, although a computer has no innate consciousness with which to perceive its belief. The unconscious belief "running" within a human brain is referred to as the universal belief (UB) that can be perceived by the innate consciousness of a human mind as the world in which that mind together with its physical attributes (real self) is living. Although real, it has much the same properties as a dream world. An actual dream is within one single human mind/brain but the dream-like world of reality is in the entire collective unconscious mind of humanity, which contains the whole UB; an individual unconscious mind can only contain an instance which is an integral part of the UB.

The UB is "created" or, more precisely, "engendered" into being by a coercive interaction between all the unconscious minds/imaginations of humanity by what amounts to a perceptual consensus. This perceptual consensus is perceived as reality, or the world in which we live. In its current form, reality is a conflicting "logical" structure because it has been engendered by individual human minds that are conflicting "logical" structures within themselves. However, when each individual becomes an integrated "logical" structure within itself the "reality" that is engendered also becomes an integrated "logical" structure, referred to as Reality, which is the total potential collective mind of humanity. The spirit of Reality "activates" that mind and is identical to the so-called Holy Spirit (God) within Christian theology at least. I use the word "engendered" rather than "created" because the spirit of Reality is critically involved in the coercive interaction that creates the UB. It works like this: The spirit of Reality potentially up-grades each conflicting human unconscious mind into a fully inte-grated "logical" structure and then "stores" a copy of it as an integral part as it becomes its own mind that is Reality. This fully integrated upgraded "logical" structure is referred to as the "Real self" of a partic-ular individual mind, as opposed to the real self it subtends in reality.

In Christian theology the Real self is referred to as the "Son of Man", who is an integral part of Reality. It is also referred to as the "Son of God", a smaller but integrated part of the whole, much like a father and son relationship in human terms. The coercive interaction can be thought of as a loving, healing partnership between the spirit of Reality and humanity (mankind). It is this relationship that allows us to have free will. One of the implications is that even God cannot pre-dict the future because there is no objective future to predict; it is engendered by the subliminal coercive interaction of the whole of humanity itself.

The coercive interaction is an ongoing interaction, the result of which is perceived as an expanding universe in the form of new dis-coveries and technological innovations. There is only here and now in Reality, which is also the case in reality, which is considered to be else-where in space and time. The past, the future and elsewhere in space are in the perception of the UB from the point of view of the here and now, implying that the so-called past is an insubstantial structure rather than concrete, objective "fact." This implies that the past is *per-ceived* to be real because we have come to collectively, unconsciously believe in its reality and not because it has been in existence from the beginning, whatever that means.

The UB is continually being "enhanced" into a lower conflicting state by certain "gifted" individual human minds that happen to be living a little closer to Reality, in some detailed aspect of reality, than the other minds around them. This engenders new technology and new "discoveries" in due course. If it so happens that a particular in-dividual human mind is living very much closer to Reality than the other minds around it, a so-called supernatural event can occur and be witnessed by all concerned. This is not considered to be a techno-logical innovation, because there is no conscious understanding of the process taking place within the collective mind involved.

Thinking in this way gives us a very different "infrastructure" in which to consider our world. It gives answers to the basic as well as the ultimate questions, unlike the space/time, atomic "infrastructure" that we use at present. This doesn't make the current infrastructure obsolete; it just puts it into a bigger picture. Facts are still facts within the new way of thinking but they are facts of unconscious belief as opposed to objective "concrete" facts, making them candidates for enhancement. This means that the so-called laws of nature can be changed by enhancement under certain circumstances.

When an individual mind is upgraded, by one means or another, into living very much closer to Reality than usual, the UB will be automatically and subliminally enhanced into a lower level of conflict by such a mind/person, and so-called supernatural events will be witnessed by all concerned. For example, the law of gravity can be enhanced temporarily in such a way that allows someone to walk on water. Again, by enhancing the mass of material objects to zero, the speed of light can be greatly surpassed. This is, in effect, so-called psycho-kinesis and as long as the level of conflict within the UB is reduced it will be effective, and the event will be creative or healing or both. If we try to deploy psycho-kinesis at the service of conscious thought the chances are that nothing will happen because the coercive interaction will intercede, as this would increase the level of conflict within the UB. This would be harmful and maybe disastrous so it would not be allowed to occur. The coercive interaction can be thought of as a protective safety net for humanity. However, a large collective mind can inadvertently enhance the UB in a detrimental way that actually increases the level of conflict within it, usually with dire consequences when the matter is redressed by the "logical" consequences of such unRealistic motivation. This is the enhancement that causes all disasters and the demise of all civilisations ruled by a unRealistically motivated collective governing process. In theological terms it provides scientific understanding of, for example, the events described in the so-called Book of Revelation.

One may well ask what the bottom line is. How can it benefit an individual human mind/being?

The coercive interaction will not intervene, as an individual mind "moves" closer to Reality by allowing the polarised mind (the conscious and unconscious) to heal and become one integrated mind that, inevitable, operates below the level of awareness, as we become our Real self. We seem to be motivated by unconscious forces; this is why such a state is referred to as living in the spirit of Reality. This is what it feels like; the most enlightened state that any life-form can attain. This could be considered to be the ultimate evolution but, as long as there is the slightest conscious involvement, there will be no change because the mind can't heal under such conditions and we stay as we are.

The Universal Belief

Current scientific understanding assumes the objective nature of what is perceived to be the universe, the world in which we are living. This means that the universe is considered to have come into existence at some time in the past, which is referred to as the "beginning". There is no argument about the reality of it. The question is: *Why* is it real? Such a question seems irrelevant if the universe is assumed to be objective fact. Now, just suppose that, instead of dogmatically adhering to this assumption, we consider that the universe is *perceived* to be real because we have collectively come to "unconsciously" believe in its reality. It is a conflicting "logical" structure in its present form, that I refer to as the "universal belief" (UB).

A belief is like software running in a computer, but a computer has no consciousness with which to perceive its belief, unlike a human mind with the "software" of the UB running within it or, more precisely, an integrated instance of the UB running within it. This is perceived by its innate consciousness as the world in which it lives. In effect, it has some of the characteristics of a "dream-like" world, although it isn't a dream because a sleeping or hypnotically induced dream is in the mind of a single human mind, whereas the UB is within the whole collective mind of humanity. The objects within the UB are of the same nature as objects within a dream; they are intrinsically imaginary objects.

The so-called Laws of Nature will operate just as well within the universal belief (UB) as within the objective universe. Facts will be the same in both interpretations but, in the first case, facts will be facts of unconscious belief instead of assumed objective facts. For example, so-called "material objects" in the objective universe are perceived to be real objects, but within the UB the same objects are actually "imaginary objects" that are perceived as solid, as possessing inertia and mass, and also considered to be factual objects. Instead, they are a fact of unconscious belief. In order to understand how "imaginary objects" can exhibit inertia, mass and feel solid requires a different understanding of the so-called electrostatic field, as stated above. Another such fact is the so-called law of gravity. Yet another fact is the speed of light, being 299,792 kps. Facts are based on axioms—self-evident truths that are assumed. We can't ask why the speed of light is 299,792 kps in a universe that is considered to be objectively real; this has to be considered a self-evident truth that we have no alternative but to accept. Within the UB, however, that question can be asked and answers found. There is a very subtle difference between the workings of the objective universe and the UB, but the implications are far-reaching.

There is no way, by studying natural events in space and time alone, to determine which of the two options we are actually living in. The same mathematics work equally well in both the UB and the objective universe. In "fact" the theory of the UB can give meaning not only to dark energy but dark matter and dark flow as well. See the chapter discussing the so-called VLA.

However, when we come to understand that the UB can be enhanced into a lower conflicting state that allows the so-called Laws of Nature to be "upgraded", under certain circumstances, before they are perceived as reality, our understanding is upgraded too. The law of gravity, for example, can be enhanced. That is, in due course it is perceived as a zero G effect, allowing certain people living close to Reality to walk on water, for example. This allows science to understand and work not only with natural events, as it currently does, but also with so-called supernatural events. Of course, under these different circumstances, such events will no longer be considered supernatural. In order to reach this state, we need a very different supporting infrastructure than the atomic space/time model that is currently deployed. We need to develop a deep understanding of the nature of the UB that is perceived as reality, how it came to be, and its ultimate potential development in what is referred to as "Reality". A metaphysical approach is necessary in order to achieve this, which means thinking outside the "box" into which we have been "educated" or, more specifically, "indoctrinated". We need to think in a scientific way, which actually means learning from Reality; effectively, a "healing" interaction with Reality. When a human mind gets closer to Reality, it starts to "heal" and, as it does, it "thinks" differently and, when it "heals" completely, it lives in the spirit of Reality. The more highly educated we become, the more difficult it is to do this, because our understanding becomes increasingly specialised and more than adequate to assure our comfortable survival in an ever more competitive world, so there is no real incentive to think differently unless one's curiosity is insatiable.

The first new concept we need to consider is referred to here as the "coercive interaction", a subliminal mind-to-mind interaction between all the minds of humanity and the spirit of Reality that "creates" reality as an ongoing process, the expanding world in which we live, by means of perceptual consensus. This is similar but more integrated than the so-called democratic interaction within the conscious domain regarding paper money, for example. Paper money has value because we all agree but if we didn't it wouldn't be worth the paper that it is printed on. It is the same with reality. This interaction creates or, more precisely, engenders the universal belief into existence by means of a subliminal perceptual consensus. It is a conflicting logical structure because it has been engendered by minds that are conflicting logical structures within themselves. It needs to be a coercive interaction in order to give it enough stability to keep it all together in the best way possible, without apparently negating free will.

The total potential collective mind of humanity, when each individual mind is a non-conflicting logical structure within itself, is referred to here as the "Real potential collective mind of humanity" or "Reality". Reality, unlike reality, is a total integrating logical structure with each uniquely individual mind, the Real self, an integrated part of it. Each Real self has a unique individuality in harmony with all other unique individuals and with Reality. This is ultimately the destiny of humanity.

From the Christian theological point of view, and thinking of the universe as a matter of objective fact, Reality is something like the Heaven that many religious people talk about, but without the religious implications. Even the ultimate human state – Reality – is engendered by humanity itself, at least in the early stages. In the meantime, too many of us are at the mercy of the coercive interaction and its conflicting nature. However, some individual minds are living a little closer to Reality than other minds in their vicinity and will automatically "enhance" the UB into a slightly lower state of conflict. Such enhancement will upgrade the UB in a small but significant detail perceived, in due course, as some kind of technical innovation or the "discovery" of a new aspect of reality such as, for example, the light photon or the atom.

Of course, those who subscribe to the objective universe consider atoms and photons etc. to be objective facts that have been in existence from the beginning, whatever that means. Such people are actually delusional, even if they are in a majority and highly qualified. Reality is not a democratic consensus, but a fully integrated logical structure. In logical terms, Reality can be considered as the Mind of God, God being the spirit of Reality, if we must use the word "God" at all.

The expansion of the universe is brought about by the continuous upgrading of the UB by means of the so-called coercive interaction and the expanding reality that it engenders. For those who consider the universe to be objective fact, the coercive interaction is perceived as the so-called dark energy that mysteriously fills the so-called "objective" universe. We need to use the word "engendered" instead of "created" here because the so-called spirit of Reality is involved in the coercive interaction's upgrading process. The "spirit" is what activates a mind in much the same way that electricity activates a computer and Reality is a mind whereas reality is a consequence of collective unconscious belief: A perceptual consensus between all human minds. It is as if humanity writes the specification for the universal belief, and the spirit of Reality confirms its existence by continuously producing an upgraded copy of reality as the potential Real universe that we are referring to as Reality. It is crucial to understand the relationship between reality and Reality in order to comprehend the various kinds of interactions that can occur between an individual human mind, which are referred to as the real self/ERS and Reality.

Another important concept that needs to be considered is the link between the real self and Reality, which we call the essence of the real self. So-called consciousness and will emanate from this entity, which also carries our status in Reality and is what some people call the "soul". This entity, of which we can never be consciously aware, becomes the Real self when an individual "moves" or upgrades into Reality. This implies that there are many other relatively similar realities into which the ERS can migrate under certain circumstances. The ERS is the only *real* thing about each individual, underlying the real self. However, even this is not entirely Real in its present state; that is why it partially lives in reality. It is also important to understand that not only is the so-called current universe contained within the UB but the "past" and "future" as well. Moreover, what has been designated the real self is also a functional part of the UB. This situation is realised when an individual identifies "his" being with the ERS instead of the real self. This implies that there is only here and now in reality as well as in Reality which, in turn, implies that all other so-called similar relative realities are in our midst. However, they are not real to us in this reality because collectively we do not unconsciously believe in their reality from here.

We could say that this understanding puts "God" back into the picture but in no way as before Charles Darwin's *Origin of Species* was written.

Gravity and the Universal Belief

To fully understand the nature of gravity, it is necessary to understand what we perceive to be the universe very differently. Instead of mistakenly assuming the universe to be a consequence of objective "concrete" fact that has been in existence from the beginning, whatever that means, it is considered here to be a consequence of collective unconscious belief, referred to as the universal belief. I can't stress this enough. Ordinarily, by observing natural events, there is no way of knowing which of the two options we are actually living in, even if we try very hard. We also need to consider those elusive entities experienced as consciousness and will.

An unconscious belief, as opposed to a thought belief, is like software running within a computer, but a computer has no consciousness with which to perceive its "given" belief. On the other hand, an instance of the "software" referred to as the universal belief running within a human brain can be perceived by the consciousness innate within the human brain. When this happens, it perceives it to be the world in which that brain is living, together with its conscious mind and its body, although it takes all the unconscious minds of humanity to contain the total universal belief (UB). This is perceived by the innate consciousness, within a individual brain, as real and solid and is in some ways similar to an individual sleeping dream world within an individual human brain/mind. In this case, however, it is not a dream as such but a dream-like world within the total collective mind of humanity in which the brain together with its body perceives itself to be living.

The UB is acquired or, more precisely speaking, engendered by means of a collective subliminal coercive interaction between all the minds of humanity. This is hopefully being enhanced, instant by instant, into a lower conflicting state than that of the current UB. However, this is not necessarily the case. Such enhancement can sometimes appear to temporarily "change" the so-called laws of nature. Of course, such enhancement could not occur in a world of "concrete" objective fact and can only occur within the UB if it ultimately reduces the level of conflict within it. Actually, the coercive interaction within this new way of thinking is the so-called dark energy that fills the universe in the conventional way of thinking.

Within the UB and the perception of it as reality, we all experience what is referred to as gravity; a mysterious force that holds things down on the surface of the earth with a certain weight. Gravity also plays a major role in holding the planets in orbits around the sun, etc. If we think of the situation existing within the UB just prior to its perception as the current reality of the universe, we can understand how the "software" running within each of our brains "places", or "pushes" objects onto the surface of the earth (within the UB) and places the planets in orbits around the sun (within the UB) according to the so-called law of gravity, but without implying the existence of a gravitational field.

It helps to think about the Starry Night software that represents a simplified model of the solar system. When it runs in a computer, it is observed as a 3D graphic display on the computer's monitor screen representing a model of the solar system of the software running in the computer. If we imagine looking at the computer screen and watching the planets orbiting the sun according to the so-called law of gravity and the laws of motion there is no actual gravity involved. If we then imagine ourselves jumping into the monitor screen and landing on the earth's surface we would experience gravity in the form of our weight on the surface of the computerised earth within the model. If, while we are in that position, we remember looking at the simulated solar system we would realise that there is no actual gravitation field within the model; there only *seems* to be because of the way the software is running within the computer. The same thing happens within the so-called UB when we look at it with the so-called ERS which is partially outside the UB.

The more elaborate "software" of the UB "running" within a human brain is perceived in due course as the world in which the human brain, together with its conscious mind and body, referred to as the real self, is living at that instant. This is similar to the way in which sleeping dream worlds are perceived within an individual human brain, although there is an important difference between a dream and the dream-like world of reality, because the whole of humanity is involved.

The enhancement mentioned above can only occur around a human mind living closer to Reality, as we define it here, than other minds involved in the same event. This situation develops what we may refer to as "Biblical charisma"; the ability to enhance the UB into a lower state of conflict. All the so-called laws of nature can be enhanced by charisma, making so-called supernatural events occur. For example, consider walking on water. A mind living closer to Reality than the other minds around it will automatically enhance the UB into a lower conflicting state in some appropriate detail, such as causing a local zero G effect by enhancement of the law of gravity which would allow the said person to walk on water if such an event was involved in a "healing" situation, which means lowering the level of conflict within the UB. Such events are considered to be supernatural because none of the minds involved realise that the UB has been enhanced or even what that means. When all of the minds involved in such events understand the dynamics of the process, all such events will be considered natural. I'm thinking of that event in the so-called New Testament where Jesus walked across the water of the Sea of Galilee to get to a boatload of his disciples just off shore from Tiberius. Jesus was living very close to Reality, in the grace of Reality if you will. Afterwards, together with his disciples in the boat, he sped across the sea to Capernaum in a instant. I will talk about this aspect of the event in the chapter on the nature of light.

As the total collective mind of humanity gets closer and closer to Reality, the appropriate "technology" will become enhanced into the universal belief, making the events mentioned above possible for a collective mind such as that of science. However, such new "technology" will probably be used in the creation of such things as artificial gravity on board spaceships, for example, rather than walking on water. When a collective mind gets very close to Reality, "technology" as we think of it today becomes redundant. Such events have already taken place sporadically throughout theological history.

How do we get to live closer to Reality, either individually or collectively? It's a matter of allowing our polarised mind, the conscious and the unconscious, to become one fully integrated mind merged with its unconscious counterpart. Under these conditions, such a mind is driven below the level of awareness, creating a feeling of being led by a benevolent "spirit"; specifically, the spirit of Reality. However, we do retain a simple child-like—but not childish—conscious mind. The Real driving force behind such a mind functions below the level of awareness in accordance with Reality. This is spirituality without any form of religion. All this is just the tip of the iceberg; a small part of an all-embracing theory that doesn't really conflict with 21st century science. On the contrary, it gives science a bigger picture to work in. It is a question of being objective and able to think outside the box into which we have been "educated". Education is indoctrination within a world of unconscious belief.

Dark Energy and The Universal Belief

To fully understand the nature of the dark energy that fills the universe, we need to understand what we perceive to be the universe differently than we currently do, instead of unwittingly assuming the universe to be a consequence of concrete objective fact, which has been in existence from the beginning, whatever that means. In this work we consider it to be a consequence of collective unconscious belief. This is referred to as the "universal belief" (UB). When the UB is perceived within a human mind by its innate consciousness, it is perceived as the world in which it lives, in the greater sense the universe; reality. Ordinarily, there is no way of knowing by observing natural events which of the two options we are living in. Even if we tried hard, everything would be perceived as real in both situations.

The UB or, more precisely, an instance of it, is perceived by the innate consciousness within a human brain as real and solid and is, in some ways, similar to an individual sleeping dream world within an individual human mind. In this case, however, it is not a dream as such but a dream-like world within the total collective mind of humanity in which the brain together with its body perceives itself to be living.

The collective unconscious belief, the (UB) is "created" or, more precisely, "engendered" by a coercive interaction between all the minds of humanity. However, when we consciously believe the universe to be the consequence of objective fact, the coercive interaction is perceived as an unidentified energy. We refer to this energy as the "Dark Energy", and it seems to work against gravity. Dark Energy is predicted along with so-called dark matter and dark flow by a set of so-called cosmological equations. Here, the equations are considered to be true, but the counterpart of the equations in the form of a visual model is currently out of date. The theory of the UB can provide an updated model, I believe!

When we "think" that we live in an objective world, we cannot appreciate that there is such a thing as a subliminal mind-to-mind interaction between all the minds of humanity. This mind-to-mind interaction actually creates reality in conjunction with Reality: the Mind of nature, if you will.

We need to understand what a universe of belief means. In doing so, it is critical that we distinguish between a conscious and an unconscious belief. A conscious belief is identical with thought, while an unconscious belief has little to do with thinking, since its roots are based in the unconscious mind.

When we are able to switch from thinking of the universe as a consequence of objective fact to thinking of it as a consequence of unconscious collective belief, it is easy to understand how a universal belief is not only able to support natural events but also so-called supernatural events. Furthermore, the UB is continuously being enhanced into a lower conflicting state, much as computer software can be continuously upgraded. The expanding nature of the coercive interaction is the cause of our perception of the expanding reality of the universe.

The universal belief is like software running within a computer. However, a computer cannot consciously perceive the results of its own "given" belief in action, whereas a human mind can perceive the results of its own instance of the unconsciously shared belief "running" within its own brain, by means of its innate consciousness. That shared unconscious belief is what we refer to as the universal belief. This is perceived in due course as the universe in which we are all living within the collective unconscious mind of humanity. It is rather like a collective dream-like world. Sleeping dreams, on the other hand, occur within individual brains as part of our individual makeup. The perception of the UB itself is not a dream because the whole of humanity is involved in its "creation". It is, in effect, created by a coercive interaction between all the minds of humanity together with the mind that is Reality. This mind-to-mind interaction is perceived furtively as the so-called Dark Energy that fills the universe when the latter is thought of as a consequence of "objective" fact.

Using this theory, we can seek answers to fundamental questions: Where does the universe come from? What is it made of? We can also give meaning not only to dark energy but to so-called dark matter and dark flow, as I discuss in the chapter concerning the so-called VLA, the Very Large Array of Dream-like worlds. I think that this is equivalent to the Multiverse when the universe is considered to be a consequence of objective fact.

Reality is of a similar nature to a dream-like world, although it is not a dream in itself because the whole of humanity is involved in its "creation". In fact, the reason why we dream is to maintain a "healthy" participation in the coercive interaction. If we can't dream for whatever reason we start to lose our grip on reality and go insane.

The so-called past is a component of the universal belief just as elsewhere in the present universe is. It can also be enhanced and such enhancement can actually change history. There is only here and now in reality as well as in Reality. To understand how "imaginary objects" that are perceived as "real objects" can be perceived as solid to the tactile senses, we need to understand the so-called electrostatic field differently than we currently do, as discussed above.

If we imagine ourselves to be in a sleeping or hypnotically-induced dream, everything in that dream will be perceived as real and solid to the tactile senses although the "laws" governing that world will be different. There may be no gravity, for example, making it possible to fly like Superman. If, in that dream, we were to ask where this world came from, we would find ourselves in the same situation as in reality, having to theorise endlessly on the issue. When we eventually wake, we will realise that the dream was within our own brain. The same applies in the "dream-like" world of reality. We need to wake from it to realise the "truth" about human existence. This is exactly what happens when we allow our polarised mind, the conscious and the unconscious, to become one integrated "logical" system operating below the level of awareness. This is exactly what all religions ultimately aspire to bring about within each of us, but we have to go beyond religion and the delusion of conscious belief to achieve it. This is the highest state of enlightenment that any life-form can attain. We then become a unique individual and an integral part of Reality.

The Nature of Light

In the Bible we read that God said "let there be light" and there was light. Whether we believe literally in that story or not it tends to fashion the way we think about light and the universe and how it came to be in the form of an objective universe which has been in existence since the beginning, whatever that means. If it read, "God said let there be consciousness and there was consciousness", it would be more to the point, because consciousness is not only the essence of light; ultimately, it *is* light. In order to understand this we need a deeper perspective than current scientific speculation can provide. I'm not suggesting that we should believe in the Biblical account of how the universe came to be.

We all perceive the world in which we live, in the greater sense the universe, as reality and all the objects that make up reality are perceived by our senses as real objects. The question we need to ask is *why* they are they real. Currently, we assume that objects in the universe are real because they are components of an objective "physical" universe that has been in existence since the beginning. Just suppose that instead of assuming this to be the case we think that objects are real because we have come to collectively "believe" in their reality and this belief is an unconscious belief that I refer to as the universal belief (UB) located within the collective unconscious mind of humanity.

When the universal belief, or more precisely an instances of it, within an individual human brain is perceived by the innate consciousness within an individual human brain, it is perceived as the world in which that individual brain, together with its body, is living.

Facts such as the speed of light being 299,792 kps and the so-called law of gravity are still facts within the UB. The difference is that facts within the UB can be enhanced by certain minds that happen to be living closer to Reality than other minds around them. The process is subliminal and automatic and this is why those involved in this kind of interaction credit God with the "supernatural" events that result, having no conscious understanding of the process taking place below the level of awareness.

I refer to a human body together with its brain etc., as the "real self" which is just another object, albeit a living object, that is perceived as real by the innate consciousness emanating from what I refer to as the "essences of the real self" (ERS) which underlies each individual real self. The ERS is what some people refer to as the "soul". The ERS is the essence of our being and when we identify our being as the ERS instead of the real self we are able to "look" at the world from a deeper perspective. This implies that the real self is also a component part of the universal belief. When we "look" at the real self together with the world through the conscious "eye" of the ERS we are able to "look" at the human eye with its retina and optic nerve connection to the ERS from a deeper perspective. There are something like 120 million photo receptors on the retina and only about 1.2 million nerve fibres to connect the signals from the retina to the consciousness of the ERS "within" the brain which means that the signals from the retina must be "multiplexed" down the optic nerve to the ERS. In effect, this arrangement slows down the "light" of direct conscious perception in the world of the ERS to the speed of light perceived by the eye of the real self. The light of direct conscious perception is actually instantaneous because it is the datum of the measurement of speed. This implies that light photons together with the speed of the photons are also part of the universal belief, both of which have also been engendered into being by certain "gifted" minds/persons that are living closer to Reality than the rest of humanity.

There may be slight differences in the value of the frequency of the multiplexing action between one human eye and another but the coercive interaction that engenders the speed of light photons into the UB deploys the value in the first instant from just one eye, the eye of the individual who is living closer to Reality than the rest of humanity at that time within the UB and uses that value which then becomes the speed of light. It will be a constant for the whole of humanity henceforth.

It could well be that the multiplexed "packets" of processed light travelling down the optic nerve to the ERS could have be taken for light photons as perceived in the world of the real self within the UB that inspired certain gifted minds to engender so-called light photons into existence within the UB.

Before Dr. Albert Einstein could discover the photon, in about 1900, it had to be engendered into the UB, as mentioned above, by a mind or minds living closer to Reality than other minds around it or them and it also had to reduce the level of conflict within the UB to get engendered into the UB at all. This could have taken place about the time of Sir Isaac Newton within the UB of that time.

From this deeper perspective, the universe of the so-called past is also a component of the universal belief (UB) which implies that space and time are also part of the universal belief. This is why I say that there is only here and now in reality as well as in the ultimate developed Reality, in which there is a zero level of conflict.

The conflicting nature of reality is a consequence of the conflicting nature of the individual unconscious minds that are "creating" or "engendering" it by perceptual consensus. The UB of the past is automatically unconsciously "enhanced" step–by-step into a lower conflicting state by "gifted" unconscious human imaginations that are living a little closer to Reality than the other minds around them.

Within the UB of the past, light photons "were" created or engendered into being real objects by certain "gifted" unconscious imaginations before they could be "discovered" and perceived as real by the rest of us in due course. This is how the whole of reality came to be "created" or, more precisely, "engendered" by humanity itself within the here and now. When we are born into reality or, more precisely, engendered into it, we inherit an appropriate instance of the existing UB, hence becoming a part of it. This process leads us to believe that we were born into the world that is reality. This whole scenario is implanted into our mind by the local coercive interaction at the point of our arrival and is just as much a fiction as the past UB of which it is part.

Faster than the Speed of Light

Being able to identify our being with the ERS is what the essence of all "religions" aspire to bring about for each one of us, as long as we do not allow ourselves to carried away by the propaganda of any religion, and amounts to the highest form of enlightenment that can be attained by any life-form. Living as the ERS will automatically enhance the UB in ways that will be perceived in due course by all those involved, without understanding, as so-called supernatural "healing" events. With understanding, the enhancement would be perceived as technological innovation. When belief enhancement changes the location of an individual real self from one place to another, it will be perceived by all those involved as faster than the speed of light travel, even instantaneous or time travel. Such events have already taken place, as recorded by various scriptures. One such is quoted in the Judeo/Christian New Testament in which Jesus of Nazareth together with a boatload of his disciples sped across the Sea of Galilee from near Tiberius to Capernaum within the blink of an eyelid. It didn't just affect those in the boat but the entire world (UB) of the day because those who walked around the Sea of Galilee to Capernaum wondered how Jesus and his disciples had got there so quickly. In this situation the so-called laws of nature such as those of motion and gravity are enhanced. The theory of the universal belief can explain such events once we are able to put to one side the mysticism and propaganda of theology.

In due course, if we should still wish to go to the stars, all it will take is for at least one member of the "space ship" crew to be living close to Reality in order to make the same thing happen to a "space ship" travelling across the "universe" i.e. within the UB. The ship will be able to travel instantly to and from anywhere in the "universe" i.e. the UB at super light speeds or even instantaneously. This is psycho-kinesis in effect, an unconscious mind-to-mind interaction and, as long as the level of conflict within the UB (reality) is reduced, it will be able to occur as a real event for all of humanity to witness as real. If we try to do it by conscious thought it will not work because the conscious mind keeps us away from Reality.

I think that before the ultimate is achieved the collective mind of humanity will steadily enhance itself closer and closer to Reality and technology-enabling super light speeds will get installed into the UB bit by bit. This is what is happening in the technological world around us as an ongoing process. However, there is nothing preventing any individual from going it alone, so long as he can overcome the indoctrinating effect of the collective conscious mind around him, in "fact" becoming like Jesus without the so-called religious implications, which were inadvertently man-made.

It would be just as easy to apparently travel through time using the same setup although we need to remember that that the past and the future are effectively an invention, one could say, a "fabrication" of human imagination. We would just be wallowing in our own self "created" (engendered) delusion.

The theory of the universal belief (UB) requires a very different infrastructure than current scientific understanding can provide. The latter is based on the conscious collective belief that the universe is a consequence of objective fact perceived as reality and that it is made up of an array of real atoms, photons and sub-atomic particles within a fabric of space and time.

The theory of the UB is based on a very different understanding of how reality came to be, as a consequence of unconscious collective belief in the form of a subliminal coercive interaction between all the minds of humanity established by perceptual consensus. The reality of the UB is a conflicting "logical" structure in its present state because it has been "created" by individual minds that are conflicting "logical" systems within themselves, although it is the best that can be achieved without negating free will. The concept of Reality is the potential Real collective mind of humanity, where each unique individual mind is a non-conflicting "logical" system within itself, referred to as the Real self.

When a new individual "real self" is "born" into the UB, an appropriate "instance" of it is acquired by the new individual via the coercive interaction as mentioned above, although the scenario of getting born into the UB is also a fiction, "engendered" by human unconscious imagination because we have come to unconsciously believe in the reality of getting born into the UB. This way of thinking is somewhat mind-bending until we become comfortable with thinking outside the box into which we have been inadvertently indoctrinated by the education system and enabled to use a metaphysical approach.

Telepathy

In order to understand "telepathy", we need to think of the world/ universe as a product of unconscious collective belief in the form of a Universal Belief (UB) as opposed to a world/universe of Objective Fact. Actually, our world is that part of the UB in which we live our daily lives, a shared belief which is established by a collective perceptual consensus in the form of a coercive interaction between all the minds in that world. This is the main theme of Dark Energy.

When the universal belief is enhanced by the mind of John, let's say, living closer to Reality than other minds around it, the universal belief becomes enhanced. Thus, the world is also enhanced, but nobody will be aware that enhancement has taken place, since it is a subliminal interaction, although with amazement they will perceive the results of the enhancement as real. Enhancement means moving the universal belief closer to Reality, which could be thought of as a social "healing" interaction. This is brought about by reducing the level of conflict within the UB by means of a kind of "electronic" tuning, there being no moral standards applied, because these belong to the realm of conscious thought.

If the universal belief is enhanced in such a way that everyone in the universe belief of that world spoke English, for example, the mind responsible for the enhancement, John in this case, would perceive everyone in the world/universal belief speaking English and would take for granted that everyone in the world/universal belief would understand what was said to them in English as well, regardless of the language others perceived themselves to be speaking. When they answered in their own language, John would hear them speaking in English. Such is the relativity of reality. Total communication would occur between all of those involved and everything would appear to be "normal", since nobody would be aware that belief enhancement had taken place and the status quo would be assumed. It would appear that John was a part of the community speaking the native language, just like everybody else. If several different languages were involved and everyone knew that, under these conditions everybody would experience a strange, maybe awesome phenomenon. I'm thinking here of an event recorded in Jewish/Christian scripture in which Peter spoke to the crowd, apparently in many different languages simultaneously, to the amazement of all those present. (Acts. 2 : 4 – 12).

If there were witnesses who knew exactly what was going on, who were completely satisfied that John could only speak English and who didn't know that enhancement had taken place, or even what it meant, and yet observed full communication between all concerned, they would be amazed. The only explanation they would be able to come up with would be that the communication was "telepathic" in nature. The word "telepathy" is a cover-up for a lack of understanding. All "supernatural" events are created in this way. Due to the lack of "correct" understanding, they are considered to be "miracles". This doesn't make a lot of sense if we continue to think of the world as objective fact. On the other hand, once we can flip our mind into thinking of the world as the UB, everything comes together, not just with so-called "telepathy" but with everything that we can experience. Once we appreciate that supernatural events are the creation of human subliminal imagination, not only the events in the Jewish/Christian scriptures but *all* scriptures become historic events in which we no longer have to believe but understand; after all we don't talk about "believing in" the battle of Waterloo, because it is considered to be historic fact.

Psycho-kinesis and Clairvoyance

There is another event in J/C scriptures that suggests "telepathic" communication. Consider the woman whom Jesus spoke to at Jacob's well in Samaria, when he told her that she had had five husbands. How could he have known that without reading her mind? The subliminal imagination is always at work exploring new ways to reduce the level of conflict within the UB. In this case (John 4 : 17 – 19) it was attempting to heal the rift between the Jews and Samaritans. Jesus, living very close to Reality, always spoke and behaved instinctively, considering it to be God's will. His actions were always healing in nature and on this occasion a social healing occurred. Jesus didn't need to read the women's mind; he just engendered the UB to contain five husbands for her whom even she would perceive to be real in her memory.

There is no objective universe to dictate who we are or where we have come from; it's all a matter of a mind-to-mind interaction, microsecond by microsecond. The only Reality is in the so-called here and now and from this standpoint everything has its beginning. Belief enhancement by suggestion (hypnosis), although intriguing, cannot achieve this effect because there is no coercive element involved. Self-hypnosis might but, if it did, it would not always be a "healing" interaction. Adverse logical consequences would be sure to follow because the "conflict level" within the UB could inadvertently be increased by self-interested motivation. This is because of the polarisation of the conscious and subconscious the human mind, both individual and collective.

If we try to use telepathy as an advanced means of communication using a conscious understanding, if it works at all, it would be motivated by self-interest that I refer to as unRealistic motivation. This would result in adverse consequences being realised because it would maintain our belief in an objective universe (please see the chapter on Disasters and Apocalyptic Events).

Darwin's Evolution by Natural Selection: As an Example

Darwin's evolution by natural selection is not really valid unless the fundamental question on which the theory depends can be answered, which is: How did inanimate matter become bestowed with life in the first instance, before the process of evolution could begin anywhere in the universe, let alone on Earth? To say that it *must* have happened is not good enough, and it's not scientific either. It's difficult to understand how such a question can be answered in the context of an objective universe. We take too many things for granted, calling them axioms without giving the matter in question any serious thought at all because it seems so obvious.

If, however, we consider the universe to be a consequence of "unconscious belief" as opposed to objective fact, precise answers can be found and some of these show that Darwin's evolution by natural selection is not what it's made out to be, and cannot Really be the means by which humanity got to be, however logical and proven it seems to be. Nor is the account given in the Book of Genesis, because that also assumes the objective nature of the universe. The theory of evolution by natural selection is a scenario within the universal belief created by the unconscious imagination of humanity, itself an attempt to explain how it came to be.

As we study this new way of thinking, we come to realise that the reality of the universe has been "created" or, more precisely, "engendered" into being by unconscious human imagination, including the series of past events and fossil artefacts in rocks that seem to prove the evolution of life-forms on this planet according to Darwin's theory.

Working backwards from the here and now, we could even say that they were fabricated into being by human imagination even if they are perceived to be real. If something is perceived as real as this is for 99.9% of us, we will be bound by its consequences, but this is not inevitable as explained below. There is a way out.

Even atoms, photons and sub-atomic particles are also engendered into existence by the "gifted" unconscious imaginations of certain individuals and do not actually exist in or very close to Reality. Bringing genetics into the discussion doesn't Really make any difference to the outcome. Before Albert Einstein could "discover" the light photon, for example, it had to be "installed" into the UB by a so-called "gifted" mind living a little closer to Reality than any other minds of humanity, maybe three hundred years before the discovery took place, perhaps by the unconscious imagination of Sir Isaac Newton or someone like him. Unconscious imaginations are always at work attempting to lower the level of conflict within the UB.

It's mind-bending to realise that, in addition to the conscious interaction between all the minds of humanity, with which we are all too familiar, there is an unconscious interaction that produces a subliminal perceptual consensus in the form of a "universal belief", an instance of what is perceived within each individual brain as the world, in the greater sense the universe, in which that individual person is living. Things in the world are real because we have collectively and unconsciously come to believe in their reality and not because they are components of an objective universe.

In order to be able to think in this new way, the current scientific understanding of the universe needs to be upgraded to contain the concept of Reality in addition to reality. The reality that is the universe in its current state is a conflicting "logical" structure because it has been engendered into the universal belief by minds that are, within themselves, conflicting "logical" structures.

When each individual human mind/being, the "real self", is upgraded into a non-conflicting logical structure, it becomes what I term the "Real self", and is the sum total of all the Real selves of humanity, engendering what I have referred to as Reality. I refer to this ongoing subliminal interaction between all human minds in conjunction with Reality as "enhancement". This interaction "creates" and continually upgrades everything that "humanity" perceives to be real. Events in the past and the future, both the natural and the so-called supernatural, are all created by this enhancing interaction. The so-called past is also part of the universal belief. This means that even past events can be changed by "enhancement" as well as the "laws of nature" such as gravity and motion. Enhancement reduces the level of conflict within the universal belief, and hence in reality too. One could refer to this as an ongoing "social healing interaction". However, the fundamental question still remains: How did inanimate material come to be bestowed with life in the first instance? In other words, what is the Genesis of life? A similar scenario applies to each of our individual "past" lives, making the problem a personal one as well: How do "I" come to be in the here and now?

To understand the answer to such a question, we need to understand with both our conscious and subconscious minds working together as one integrated "logical" unit and to realise that the most integrated "knowledge" that we can ever have is the "logic" contained within the Real self, being only an integral part of the truth that is Reality and not the whole of it. Furthermore, the Real self is totally below the level of awareness so we can never have a conscious understanding of that "truth". The Real self is what a human mind becomes when the polarised mind – the conscious and unconscious – "heals" into one integrated "logical" mind. It is also a state of being, when living in the spirit of Reality or, in Christian theological terms, living in the Holy Spirit. It is important to realise that this state of affairs has very little to do with any religion. The subject of religion will be dealt with in upcoming chapters. This means that we can never have an integrated conscious understanding of the truth, meaning and purpose of our lives or the world in which we are living but we *can* live in the truth by living in our Real selves as an integral part of Reality. Eternal life doesn't mean living forever somewhere after the "death" of the real self; it means living outside space and time in the so-called here and now. This is a state of being unRealised by most of us.

As the level of conflict within the UB – reality – approaches zero reality (Reality), this can apply to an individual or a collective state of being. This is what I refer to as "living in the spirit of Reality" and it is the highest state of enlightenment and the highest state of being that can be attained by any life-form. In theological terms, it is identical to living in the Holy Spirit. The Real self is what Jesus referred to as the "Son of Man". This is a joyful state of living in the here and now, as opposed to living in a vacuum between yesterday and tomorrow as most of us currently do. This is exactly what the one Jesus of Nazareth is trying to get across, once we can interpret His words Realistically. As mentioned above, this state has little to do with any religion. Instead, it is a matter of becoming a unique individual living in harmony with others as an integral part of Reality.

At first sight, this is a mind-bending concept, because it entails thinking outside the so-called "box" into which we have all effectively, if inadvertently, been indoctrinated by the "education" system.

The above is a small but integral part of the much bigger picture that I refer to as The Theory of Universal Belief.

Darwin's Evolution by Natural Selection: The Bigger Picture

Darwin's evolution by natural selection is not really valid unless it, or some other means, can answer the fundamental question on which the theory depends and this is: How did inanimate matter become bestowed with life in the first instance, before the process of evolution could even begin anywhere in the universe, let alone on Earth? To say that it *must* have happened is not good enough and it's not scientific, either. It's difficult to understand how such a question can be answered in the context of an objective universe. We take too many things for granted, calling them axioms without giving the particular matter in question any serious thought at all because it seems so obvious, although we wonder!

We unwittingly assume the world in which we live, in the greater sense the universe, to be a consequence of objective fact, meaning that it came into being sometime in the past and has nothing to do with us, humanity. This is not something that we are taught; it is something that is assumed because it seems to be so obvious; what else could it be but objective fact? As a consequence we find it intriguing and puzzling that we are unable to find satisfactory answers to such basic questions as:

1) How did the universe come to be?

2) What is it made of?

3) How did inanimate matter get bestowed with life before evolution could even begin anywhere in the universe, let alone on the planet Earth?

To get answers, we need a different approach other than assuming our world to be a matter of objective fact. This is just what I try to spell out briefly below, using Darwin's evolution by natural selection as an example. When we come to think in this new way, however mind-bending it may seem to begin with, we can understand that Darwin's so-called evolution by natural selection is not what it's made out to be and is not the way that humanity came to be on the planet Earth. It's not just a matter of shuffling organic molecules around until they acquire life, because life is consciousness and will that emanate from the so-called ERS (see above) that has little to do with space and time. Even bacteria has a real self and an ERS, but if we think that bacteria was bestowed with life something like four billion years ago we are back to the Book of Genesis all over again, just four billion years earlier where God, the spirit of Reality, bestowed an ERS into a real self which was a group of organic molecules and started life on this planet. This is still assuming an objective universe and it still leaves us asking how the universe came to be. Whatever we come up with, we will always need to ask, "But where did *that* come from?" and so on, unless we say that God, the spirit of Reality, did it by "magic" some fourteen billion years ago. To break with this kind of scenario we need a very different approach.

There is a better way of thinking about how life was bestowed into inanimate matter in the first instance once we think about a universal belief as opposed to a universe of objective fact.

The description of how everything came to be in the so-called Book of Genesis is not the way we came to be either because that also assumes the objective nature of the universe as stated above. Even if we move the timeframe back four billion years and consider "Adam and Eve" to be those first primitive life-forms, bacteria. This scenario is not able to provide a more satisfactory answer for the Genesis of life than the Adam and Eve story itself! The answers given to the above question are intriguingly different, even devastating, if the theory of the so-called universal belief turns out to be true.

Here, we are going to consider Darwin's Evolution by Natural Selection from the point of view of a universe of unconscious belief as opposed to a universe of objective fact, which is the underlying theme of *Dark Energy*. According to this way of thinking, objects are real not because they are component parts of a physical universe of objective fact but because we have come to collectively believe in their reality. This belief is a belief within unconscious rather than conscious minds. This is not the way we normally think about the word "belief", as we have become accustomed to using the word "belief" to imply thought. When we say "I believe", we mean that we really think a statement or something to be true in the sense of factual reality. It will not be easy to think in this unconventional way until current scientific understanding has been upgraded to include so-called supernatural as well as natural events.

The sequence of artefacts in the form of fossil remains in rocks that are used as confirmation of the reality of Darwin's theory of evolution by natural selection are considered here to exist in the past part of what is being referred to as the universal belief (UB) rather than the past universe of objective fact, perceived to be the reality of the past. The sequence is "logical" because human unconscious imagination has made it so, starting from the here and now, by "engendering" ancient artefacts into the "past part" of the UB, little by little, prior to their "discovery" as real archaeological artefacts and events in the here and now of the so-called 21st century.

It is too easy to fail to appreciate that there is an unconscious interaction between all the minds of humanity. This interaction "creates" or, more precisely, "engenders" all aspects of reality into existence, a process of "enhancement" that is ongoing. From the cosmological point of view, it is perceived to be the expanding "universe" and it shows itself furtively as that mysterious dark energy that fills the "universe." Here, this is referred to as the coercive interaction, a subliminal interaction rather than the more familiar conscious interaction between all the minds of humanity.

Even the so-called radioactive elements within the rocks that are used for dating the fossils and as confirmation of the time scale of the various life-forms found within the "past" part of the UB are engendered into the UB "prior" to their "discovery" and are subsequently also perceived as real objects in the form of electrons, photons and sub-atomic particles etc., within the UB. Even space/time itself is engendered by the coercive interaction as a "subliminal perceptual consensus" into the UB. The only factual reality is in the so-called here and now.

Both elements of the human mind, conscious and unconscious, are "pattern" seeking devices by means of their innate imagination. Certain individual subliminal imaginations closer to Reality (see below) than the rest of humanity continuously come up with new "ideas". If such new "ideas" can reduce the level of conflict within the UB they are "installed" automatically into the UB via the so-called subliminal coercive interaction between all the minds of humanity. In due course, the newly "installed" components of the UB are "discovered" by conscious minds that are continually searching for a conscious integrated pattern.

Reality with a capital "R" is the total potential collective mind of humanity when each individual mind is an integrated logical structure within itself, unlike the human minds that engender reality which is, consequentially, a conflicting logical structure. Such an individual mind is referred to as the Real self as opposed to the real self; "physical attributes", in reality. One may well ask how God, the spirit of Reality, becomes involved in the creation of the UB. The spirit of Reality can be thought of as a loving, healing spirit that takes each human mind in its present state and upgrades it potentially into its Real self, retaining a copy of it as an integral part of what becomes its own Mind – Reality.

Individual human minds can only enhance the UB into a lower conflicting state as a consequence of this upgrading process. The whole of humanity is continually upgrading itself toward Reality. We could say that the spirit of Reality creates the UB and hence the universe to human specifications; this is how we come to have free will. Although "driven" by the spirit of Reality, it is in effect a partnership between the spirit of Reality (God) and mankind by means of an interaction of minds.

A collective human mind can, however, "drive" the UB into a more conflicting state by becoming more and more unRealistically motivated. In this case, it is only a matter of time before the coercive interaction itself creates the future as the "logical" consequences of the unRealistic motivation which will usually, but not always, be adverse. UnRealistic motivation can take the form of greed or grief, for example, and unRealistic motivation eventually dooms all civilisations driven by conscious thought..

Michele Nostradamus and the Future

So far as the future is concerned, it cannot be predicted because there is no objective future to predict. Michele Nostradamus, a man with a so-called "second mind" (see below) was able to enhance the UB by installing dramatic future events which could be perceived as real events in due course, even after his death. It seemed as though he was able to predict future events. He was a man with considerable charisma as a consequence of having a gifted relationship with the spirit of Reality in the form of a second mind within the other half of his own brain. This second mind was living much closer to Reality than his own mind or that of those around him, and was therefore able to enhance the coercive interaction and install future events into the UB.

This is the same interaction as that of a creative single mind when it installs the photon, for example, into the UB. In the case of Nostradamus, it was done with more dramatic flair. This effect is merged with the creative efforts of single minds as well as the logical consequences of unRealistic motivation of collective minds, as mentioned above, so that the "prediction" becomes defocused. We will discuss the Nostradamus effect further in the next chapter.

We have discussed a brief description of the very different infrastructure being used in the theory of the UB, as opposed to the space/time/atomic infrastructure that is in sole usage at the current time. Reality is outside space and time between the space/time of yesterday and tomorrow, where we all should be as our ERS. Instead, we live in a vacuum between yesterday and tomorrow, lost in what we call space and time with little or no presence in Reality. The process described in the paragraphs above seems to confirm that the evolution of life on earth started about 4 billion years age as primitive life-forms and then evolved, together with the earth, into so-called modern humans. The ultimate question in this context is: How was inanimate matter bestowed with life before the process of evolution could even begin? This is one of the fundamental questions and I maintain that it cannot be answered satisfactorily in the context of a universe of objective fact, i.e. by assuming the universe to have come into existence in an objective past.

It is assumed that it must have done but this is not the kind of answer that is really scientifically good enough, let alone acceptable.

This means that Darwin's so-called evolution by natural selection is *not* the way humanity came to be in the world today, and nor is the account in the Book of Genesis, because that too assumes the universe to be a consequence of objective fact.

What I assert here is that "factual" reality is only in the here and now; the answer to the question of how life was bestowed on inanimate matter has to do with the human consciousness that is the "centre" of our being as a life-form in the here and now. To understand this, we must understand with both our conscious and unconscious minds working together as one mind. This can only be achieved by allowing them to integrate into one "logical" mind which will, inevitably, operate below the level of awareness. The answer cannot be understood by the conscious mind alone, implying that we can never have a conscious understanding of the "truth" of human existence.

As stated above, we don't seem to realise that all of those so-called archaeological "facts" perceived as fossil records, etc. have been engendered into existence by certain individuals minds living a little closer to Reality than those around them via the subliminal coercive interaction, before they can be perceived as real events within that part of the UB we refer to as the past. Those individual minds automatically enhance the UB, even the "past" UB, by enhancing or upgrading a particular aspect of it in such a way that reduces the level of conflict within it. In due course, these are perceived and consciously believed or understood to be a newly discovered "fact" in archaeology, science or technology. The new "fact" is said to "prove" the truth of whatever new conscious belief theory is being proposed.

We are all involved in various aspects of this interaction one way or another which means that there is a continual subliminal "communication" guided by Reality at the same time as allowing free will, between all the minds of humanity; the so-called coercive interaction. This interaction even "engenders" the space and time in which all such activities occur. In "fact", every aspect of our intriguing world has been engendered, in effect created, by the subliminal imagination, as a perceptual consensus by humanity operating in the here and now.

From the geological point of view, we could also ask how the earth on which we live came to be just right for life as we know it today. The answer is that it is just right because we have *made* it so with our unconscious imaginations "engendering" it into "existence" within the "past" component of the UB.

There are those who consciously believe that extra-terrestrial life-forms "put" life on this planet sometime in the "past". Please bear in mind what I am saying about the so-called past. We still need to ask how inanimate matter was bestowed with life in order to produce life "elsewhere" in the "universe," which is actually elsewhere within the UB.

Like the current part of the UB, the past part of the UB has been "engendered" into existence by a human subliminal mind-to-mind interaction in the here and now. All of the so-called archaeological discoveries which appear to prove current understanding of "past" events explain the evolution of life in this world we call reality. This is what we would expect. The starting point is here and now, so every-thing has been tailored into the UB of the past to be consistent with the here and now. However, the fundamental question of the origin of life still remains.

The answer to this question revolves around the "living fact" that, in the here and now, I, like the reader, am bestowed with conscious-ness and will (life) as a "fact" of experience. Each of us needs to ask: "How did *I* come to be in the so-called here and now?" This is the only question that needs to be answered. The response is a "living" answer as well as an intellectual one. To "truly" understand the answer to such a question, we need to realise that the most integrated "knowl-edge" that we can ever have is the "logic" contained within the Real self, which is only an integral part of the truth that is Reality and not the whole of it. Furthermore, the Real self is totally below the level of awareness. The Real self is what a human mind becomes when the polarised mind – the conscious and unconscious – "heals" into one integrated "logical" mind. It is also a state of being when living in the sprit of Reality or, in theological terms, it is living in the Holy Spirit. It's a matter of extracting our consciousness, our soul, or as I call it, the ERS, from its addiction with the conscious mind with its knowledge and understanding. Otherwise, it becomes trapped, making it impos-sible to live in the here and now. Under these conditions, we identify our being with the ERS instead of the real self. This means that we can never have or need an integrated conscious understanding of the truth, meaning and purpose of our lives but we *can* live in the truth by living in our Real self as an integral part of Reality. Eternal life doesn't mean living forever somewhere; it means living outside space and time in the so-called here and now. As the level of conflict within the UB and hence reality approaches zero reality, Reality, this can apply to an individual or a collective state of being. I refer to this as "living in the spirit of Reality" which is the highest state of enlightenment, or rather non-enlightenment, and the highest state of being that can be attained by any life-form. In theological terms, it is like living in the Holy Spirit, which is exactly what the one Jesus of Nazareth is trying to get across, once we can interpret His words Realistically. In effect, it is the so-called salvation that Jesus represents and that some rave about.

The Future and its Prediction: The so-called Nostradamus Effect

So far as the future is concerned, it cannot be predicted because there is no objective future to predict. Michele Nostradamus, for example, was a man with a "second mind" (discussed in more detail below) and consequently was able to enhance the UB by installing dramatic future events which could be perceived as real events in due course, even events meant to occur after his death, making it appear as though he was able to predict future events. He was a man with considerable charisma, as a consequence of having a gifted relationship which the spirit of Reality in the form of a second mind within the other half of his own brain. Because of its simplicity, this second mind was living much closer to Reality than his own mind or the minds of those around him and was thus able to enhance the coercive interaction and install "future" events into the UB.

Nostradamus is not the only voice predicting the so-called future. An array of ancient cultures have done the same thing within the universal belief, and many seem to concur that an apocalyptic event is scheduled to occur in the year 2012. The reason that such a diverse array of so-called prophecies seem to converge in 2012 is because we, in the here and now, have come to unconsciously believe in such a scenario, making it seem as though ancient cultures had some Godly common wisdom. If we (humanity) have really come to unconsciously believe in such a scenario it really *will* occur in 2012. The polarised minds of humanity in its current form dream up such apocalyptic events. The remedy is simply to allow our polarised minds, the conscious and unconscious, to merge into one integrated mind structure.

This single, integrated mind will inevitably operate below the level of awareness in each human mind, giving us the feeling of being driven by unconscious forces as unique, individual minds living in perfect harmony with each other with each other as integral parts of Reality.

Precognition

This is the same interaction as that of a creative "gifted" single mind when it installs the light photon, for example, into to the UB. In the case of Nostradamus, it was done with more dramatic flair, because of the charisma of his second mind. This effect gets merged with other charismatic "predictions" by other second minds together with the creative efforts of "gifted" single minds as well as the logical consequences of unRealistic motivation of collective minds, as mentioned above, so that the "prediction" becomes defocused.

There were, of course, many others with the same gifted relationship with the spirit of Reality together with all the other various interactions mentioned above, making up what are sometimes perceived as real future events in due course. We could describe this as a conglomerate, ongoing mix of charismatic enhancement.

Jesus of Nazareth was another very charismatic historical figure who did the same thing with his own life, although his charisma was due to do a direct relationship with the spirit of Reality rather then a gifted second-minded relationship. None were actually predicting events in an objective future as a form of so-called precognition. His resurrection was installed into the UB of the future by Jesus himself before he died, making it appear that he did actually come back to life. Indeed, those events were very real. But however real that scenario seemed to be to all those who witnessed it, it was all just another invention of the human imagination; in this case, of Jesus himself. He didn't arrange any of those events by any conscious plan; they simply occurred because they reduced the level of conflict within the UB.

The resurrection of Jesus is a vital part of Christian theology and charismatic fervour and it gives us the impression that we shall rise from our own 'death' and appear in 'heaven' with the appearance that we have in reality but this is not the significance of this event. The significance of the event is the direct interaction with Reality which is a kind of evolution of the ERS out of reality into Reality in one migration instead of migrating from one reality to another until we arrive in Reality.

The reason why Jesus' body was never found is because before he 'died' he enhanced a negative hallucination to that effect into the future UB inadvertently of course, meaning that there was no conscious involvement, he always credited God with all the social healing events that happened around him. That caused his body to ceased to exist within the then current reality. It is the same phenomenon that causes things and people to disappear without trace within the so-called Bermuda triangle. Please see below.

Time Travel

When a person with a DTH profile and "appropriate" second-minded charisma (see note A, below) hallucinates, either spontaneously or by intent, a "vision" of a potential past or future event, that event will be automatically enhanced into the UB. This makes it real, providing it reduces the level of conflict within the UB. Such future events are perceived as real events in due course. When past events are hallucinated, they are also enhanced into the UB of the past in the form of artefacts, depending on the time scale, of either historic records or even hardware or fossil records, in one form or another. This actually changes or updates history. A group of people could also get involved in much the same way as UFO's are perceived as real, as we will explore in detail later. We need to bear in mind that such things can only become real if they reduce the level of conflict within the UB.

When a person with a DTH profile and appropriate second-minded (SM) charisma, hallucinates a vision of "himself" within a potential future or past event the event is real for him. If the "vision" was also able to reduce the level of conflict within the UB it would be automatically be enhanced into the UB as a real event for all involved. It would seem that he had travelled through time, not to himself but to all those involved in current and specific future and past events. He would be able to "prove" that he had been into a past or future event because future events would become real in due course and past events would show themselves as ancient artefacts in one form or another with our time traveller as a component part within the past part of the UB. If there were witnesses who ardently believed that the world in which they were living was an objective one, they would witness time travel and have no option but to think of it as a supernatural event or some kind of mass hallucination, even if they wondered! As I have already said, such events can only become real if the level of conflict within the UB is reduced. If any form of conscious aspiration is involved, it is impossible, because self-interest is inevitably evoked and this is more likely to increase the level of conflict within the UB than to reduce it.

Only certain people who have given up conscious aspiration can "create" or, rather, engender such events into the UB. One example is the one Jesus of Nazareth, a man with a second mind inherited from his mother (and therefore the lineage that has been traced back to King David's father, Jesse). The so-called John the Baptist, who also had a second mind, was able to upgrade Jesus into a direct interaction with Reality, making him one of the most and maybe even *the* most charismatic figure in history. One incident that took place around Jesus was his so-called transfiguration; a good example of apparent time travel. For more details regarding this, please go to Wikipedia: The Transfiguration of Jesus.

Even so, such events would in truth (Reality) all be a consequence of human subliminal imagination as is the rest of reality that we call the universe. There is only here and now in both reality and Reality.

Note A on the meaning of appropriate second minded charisma.

Second-mindedness (SM) is a family trait that affects about one in twenty humans. We can say this because SM goes hand-in-hand with another family trait; that of being hypo-sensitive to hypnosis, which means having a so-called Deep Trance Hypnotic (DTH) profile; the ability to spontaneously hallucinate anything that one's mind becomes fascinated with. The two family traits always go together.

The SM/DTH family trait affects different people in different ways. Some are not aware that they have SM/DTH profiles whereas others become wonderful spiritual healers and attract large followings.

Many become very religious, some secretly so because they perceive their SM to be God on account of its "supernatural" powers. This is what drives the world's religions. Living very much closer to Reality than humankind, the SM is consequently able to enhance the UB into "miraculous" healing events and sometimes into "miraculous" hurtful events referred to as "the wrath of God". This is because second minds are life forms in their own right even if they are living closer to Reality than humankind. They are just as concerned as human life forms with their own survival, and will react just as humans do against life-threatening situations, even if it means injuring or killing the cause of the threat. They don't have arms and legs, as people do, but they can use what amounts to psycho-kinesis which will usually reduce the level of conflict within the UB, even if the event looks gruesome from the humans' point of view.

However, there are some who can be more objective about their "gifts". These people are surrounded by supernatural happenings, usually in the form of healing events on a grand scale, that can also involve "apparent" time travel and the like as in the case of the so-called Transfiguration of Jesus, mentioned above.

The world in which we live is far more complex than any of us can ever imagine!

UFOs, Extra-Terrestrial Life Forms and Religion

To understand UFOs, we need to realise that not only is the universe a consequence of collective belief, termed the Universal Belief or UB, but also that one in twenty individual human beings are "deep trance hypnotic subjects", which means that they are very susceptible to hypnotic suggestion, and can spontaneously hallucinate anything that their mind becomes fascinated with. This is a gift because the individual is born with it as a family trait. Furthermore, any sort of person can fall within this category. This means that one in twenty doctors, airline pilots, clergymen, school teachers, policemen, young or old, male or female etc., etc., have DTH profiles. There is no way of knowing who has and who does not have a DTH profile other than by hypnotising them to determine their hypnotic sensitivity. There are some 300 million individuals on the planet Earth with DTH profiles, so it is not surprising that there are numerous UFO sightings. These will remain hallucinations until the Universal Belief (UB) contains them, when UFOs will become real, and someone will be credited for "discovering" them or making first contact.

However, another family trait goes hand-in-hand with DTH profiles and this is second- or double-mindedness (see later paragraphs). The two traits work together, making it possible for an hallucination to be transferred into the "local" UB by the charisma emanating from a second mind. This is how Moses (a man with a second mind) could separate the waters of the Red Sea. These are examples of positive hallucinations being installed into the UB by so-called second-minded (SM) charisma.

A Dubious UFO Sighting?

A friend of mine was travelling with his family from the Isle of to Wight to Norfolk. They were travelling in an RV and when they arrived in the valley of Ely they decided to stop and rest for the night. During the night, my friend awoke and went outside. He was amazed to see lights moving in straight lines in the sky, sometimes making right-angled turns and seeming to turn on and off. When the event was related to me later, I wondered what it could have been. I soon came to the conclusion that the lights were not UFOs but rather *UFIs*, un-identified flying insects, rather like the fire flies in the United States. They were flying about three metres above my friend's head and there were so many going on and off all the time that some of them created the illusion of making right-angled square turns, which would be against the laws of motion for a material object. What was really happening was that two insects were flying at right angles, and one turned on at the same point that another turned off, giving the im-pression that it was the same one doing a right-angled square turn. Of course, my friend stood by his conviction that he had witnessed a UFO sighting. This just goes to show how careful we need to be when observing an unusual phenomenon that might turn out to be a "gen-uine" UFO sighting.

The so-called Bermuda and Dragon Triangles

On the other hand, when negative hallucinations get into the UB, things and people can disappear without trace. The events that occur in the so-called Bermuda Triangle are an example. Consider the crew of the sailing ship Mary Celeste, that disappeared without trace. For further information please look up Mary Celeste in Wikipedia. One of the crew members presumably had a second mind in addition to a DTH profile with devastating consequences. Again, the cause is cha-risma emanating from second minds coupled with a DTH profile. This is actually the same phenomenon as that which causes UFO sight-ings. A negative hallucination occurs when a hypnotised subject is told that a certain object or person within a group that everyone else within the group perceives as real does not exist, so that the subject will not see that object or person. Instead, he will hallucinate the background behind the now "non-existent" person. If this event is conducted in the presence of an SM person who can project the negative hallucination into the UB, no one within the group will see the "now non-existent"; person, he would just seem to have disappeared or to have never existed and the rest of the group members wouldn't remember him either! However, this could only happen if the level of conflict within the UB was reduced, so it would be a most unlikely, if not impossible, scenario. Events such as these happen spontaneously in areas like the so-called Bermuda triangle from time to time.

One may well ask what happens to people who disappear in this way. As the reader will understand as we progress, when the real self no longer supports the ERS, by what ever means deployed, the ERS will migrate and be integrated into another reality within the VLA (see below) acquiring a new, appropriate real self, depending on its status in Reality. The memory of the "non-existent" person or object within the minds of all the group members will also cease to exist. Incidentally, the disappearance of the "physical" body, the "real self" of Jesus, can be explained in the same way, although in this case he did it himself before he "died". This effect will be discussed in more detail in the chapter concerning the VLA. There are many "shades" to this phenomenon so that many types of events will be engendered into reality this way, although on occasion there may be adverse logical consequences for all those involved within the group and maybe for the whole of humanity if global unRealistic motivation is inadvertently deployed.

The difference between UFOs and technological innovation is due to the type of imagination deployed. In the case of UFOs, imaginations that lack integrity are involved whereas in the case of technological innovation, better-integrated imaginations are involved. The UB is enhanced in this case, and not so much in the former, without the need for SM involvement. When the UB is enhanced, "discovery" follows. In the case of technological innovation, this is immediate. It is not considered to be a "discovery". Instead, it is creating something real from an idea within an integrated subliminal imagination, integrated with the UB, that will in due course be "discovered" as real.

Belief enhancement by suggestion (hypnosis) usually has no coercive element, as does enhancement by the coercive interaction between minds, unless second mind charisma becomes involved. UFOs will not be real objects unless the human collective subliminal imaginations can incorporate the necessary logical structure into the UB to make them real in due course. The same applies to "extra-terrestrial" life-forms. Human subliminal imaginations "create" what we perceive as the universe, including extra-terrestrial life-forms if these exist. This universe incorporates the world in which we perceive ourselves living, as opposed to the world of objective fact that we have come to take for granted.

Where Angels Fear to Tread acquired its title because there is one place where even angels cannot go and that is Reality. Angels don't fear to go there, they just can't, because angels result from human subliminal imagination with perhaps a touch of SM charisma, just like UFOs and extra-terrestrial life-forms. Even our own "physical" bodies, referred to as the "real self", are products of collective belief. For UFOs to be considered real, they must be perceived as real by the whole of humanity. The same is not always true for angels. An hallucination of an angel is real to the hallucinating person but no one else. Especially in combination with second or double-mindedness, it is indeed awesome. To Mohammed, the Angel Gabriel was awesomely real, so much so that a new religion, Islam, was born. The psycho-dynamic nature of second-mindedness affects about one in twenty, although not usually as deeply as it did the so-called prophets. This is the phenomenon which produced the Old Testament prophets and many more. Second minds are life-forms in their own right, living closer to Reality than any human mind and acting as spiritual hosts to their "physical hosts", the second-minded people. This is the realm of the Gods.

Egyptian and other Pyramids

The Egyptian, Greek and Roman gods, for example, were a consequence of SM charisma and may be the reason why such civilisations were able to move the heavy stone objects used in their building projects such as the Egyptian Pyramids. They would have been able to use levitation to help lift such stones into place! Some think that extra-terrestrial life-forms could have been involved, but there is little difference between intermediary seconded minded "gods" and extra-terrestrial life-forms since they would have been living much closer to Reality than humanity. It's an intriguing thought!

When we think of other pyramids around the world and wonder if there could be a connection between the various civilisations that built them, it seems unlikely, as they are so far in the "past". But then we come to understand that we are not living in an objective universe and that human unconscious imagination has created the so-called past; it's not surprising that there is a similarity between pyramids in Egypt and the Americas. For more information please look up Mesoamerican pyramids in Wikipedia. But I digress!

When second-mindedness is combined with a DTH profile and the two act together, sparks will and do fly. The charisma surrounding SM hallucinations can get into the UB because it reduces the level of conflict with it. This is, in effect, social healing in and it produces what bystanders will perceive as miracles on a grand scale. Think of Moses and Joshua. Moses hallucinated the parting of the Red Sea. This was immediately transferred into the UB by the charisma of his second mind and so it became real for everyone, which was awesome indeed.

A similar thing happened to Joshua, when the sun remained stationary in the sky for one whole day. Everybody involved assumed that God was responsible and, in a way, they were right, since the second mind of Moses and Joshua, posing as God (Jehovah) was able to enhance the UB. Today, many people think of God as a super-being who exists somewhere and who can do such things. Here, we understand "God" to be the Spirit of Reality, referred to as the Holy Spirit in certain theological circles.

It is difficult to retain a strict focus on UFOs, because everything is part of an integrated picture, like a jigsaw puzzle. The human subliminal imagination becomes an integral part of Reality as we become our Real self. The coercive interaction which creates the UB becomes redundant, as do DTH profiles. If there are angels in Reality they will be as Real men or women; integral components of Reality. We all become small integral units of Reality, smaller parts of the whole. From an emotional viewpoint, this is like a father and son relationship. Since Reality is likened to the mind of God, we all become sons and daughters of God when we dwell in Reality.

If a large collective mind (group) starts believing in a different way then the rest of humanity, such a distorted belief can engender an abnormal coercive interaction that will actually increase the level of conflict within the UB and hence what is perceived as reality. We could say that such a group becomes unRealistically motivated. In due course, the normal coercive interaction will normalise the situation by creating the logical consequences of the unRealistic motivation which will be perceived as adverse in some appropriate way, just in case of Noah and the flood.

The individuals within a group that become abnormally or unRealistically motivated do not realise what is happening to them because they support each other in what amounts to a delusion within the world delusion. For example, greed is not recognised as such when everyone in motivated similarly, and we must not forget that even the so-called normal coercive interaction engenders a delusion that is perceived as the reality in which we live when viewed from the point of view of Reality. This delusion, believing the universe to be a consequence of objective fact, is the reason why we in science cannot find the answers to fundamental questions such as: How did the universe come to be? What is it made of? How did inanimate matter become bestowed with life before evolution could begin anywhere?

A group that believes that there are "real" extraterrestrial life-forms living among us will be rudely awakened by the logical consequences that will result. Such consequences could be perceived as humanity being taken over by a superior life-form, for example. Such a state of affairs would be "engendered" into reality by humanity itself as a result of what amounts to unRealistic motivation. There are no such things as unsolicited accidents and disasters; they all result from the "logical" consequences of unRealistic motivation and this is what causes the demise of all civilisations sooner or later.

We need to take into account the so-called coercive interaction between all human minds to be in a position to avoid the involvement in unRealistic motivation, although the results are not always catastrophic. This depends on the type of unRealistic motivation involved, such as greed or grief, for example. In the global sense, this is the essence of the so-called Book of Revelation: The ultimate catastrophe.

On the other hand, when the so-called apostles of Jesus were dev-astated by the death of Jesus they were unRealistically motivated by grief. The logical consequences were totally joyful and enlightening, so much so that they were said to have received the so-called Holy Spirit *Dark Energy* gives an integrated understanding of everything humans can experience, both natural and supernatural. Just like a jig-saw puzzle, we can't see the whole picture until every piece is in the correct place.

As mentioned above, about one in twenty individuals are born with and "gripped" by a family trait in the form of a second individ-ual mind within the other hemisphere of their brain. Because of its simplicity, that second individual mind dwells closer to Reality than the host mind. This mind-to-second mind combination drives all the world's so-called religions. The simplicity of the second mind results from its not having to earn a living or bring up a family, etc., so it can maintain a greater propinquity with Reality than that of the mind of its physical host and is therefore able to "engender" the UB into a lower state of conflict under certain circumstances, hence promot-ing so-called supernatural events. The second mind is sustained by its host much as children are sustained by their parents. Those few "gifted" people who have a so-called second mind within their own brain perceive it from a deeper perspective, because they tend to identify their being with the ERS instead of the real self. What they perceive is a second "divine" ERS outside themselves, which they con-sider to be God. In a way this is true, because behind that "divine" ERS is the spirit of Reality which is the Real and only God, if we must use the word "God"' at all. In fact, it would be more accurate to consider the second mind to be an intermediary between the spirit of Reality and the individual person. This is the interaction we read about in the so-called Old Testament, for example, that can create so-called super-natural events. Such events are referred to as supernatural because the other minds involved are unaware that enhancement has taken place or even what that means. They simply perceive unusual and sometimes frightening events, which can be misinterpreted.

The birth of Jesus was a supernatural event, within the UB of the past that was engendered by the "gifted" unconscious minds of the so-called Prophets, within the Jewish Nation's collective unconscious mind that in due course was perceived as a "virgin birth". This event upgraded the so-called Old Covenant interaction with Reality to the so-called New Covenant interaction, a direct interaction overcoming the shortcomings of the second-minded intermediary interaction. As far as Christians are concerned, the intermediary is sometimes perceived as Jesus or what the host person believes, in "his" heart, to be Jesus. It is also considered to be the God that has no name and of course this is exactly what Jesus is; an intermediary between an individual human and the spirit of Reality. This must, of course, be awesome and very personal to the host person because supernatural events can occur.

This kind of interaction with the spirit of Reality can only be realised by those "gifted" one in twenty; the rest of us are expected to believe what they tell us about "their" God. This resulted in what we call faith and of course there are benefits to having faith, but this is not what Jesus is really talking about, although he could have had a second mind, inherited from his mother and hence from the line going back to King David. This is why Jesus is often referred to as a son of David. In fact, under certain circumstances, the second mind can be concerned for "his" own well-being even its continued existence and can induce, by "negative" charisma, hurtful effects within the UB directed at persons who seem to pose a threat to "his" well-being. Such people experience hurtful consequences or even death as in the case of the man and wife who sold a piece of land and gave only part of the proceeds to St. Peter, as described in the Acts of the Apostles. When charisma is deployed in this way it is said to put a spell on the targeted person. Thus, the UB is enhanced detrimentally against the targeted person. Even the host person can be subject of negative charisma produced by a "concerned" second mind, as in the case of Job.

Charles Darwin s Illness

Charles Darwin, as another example, was a very sick man. He had a mysterious mental condition, possibly caused by the charisma coming from second-minded persons near him, because he went up against religion, hence the Gods, which means those second minds around him would have experienced concern. More than one second mind may have been concerned with what he was doing. The Bishop of Oxford, Samuel Wilberforce, could have been one of the culprits, unawares. Even Charles's wife, Emma, who was a very devout Christian, could have been another, even without realising it.

Sometimes I wonder if something like this is happening to me, although I'm not going up against traditional religion or the Gods as strongly as Darwin did. In fact, I'm not trying to discredit the Gods or religion at all; just the opposite. This work puts both into one overall understanding that is referred to as the Universal Belief. It actually puts God back into our lives in science, as the spirit of Reality and in no way as "it" was before the publication of *The Origin of the Species* by Charles Darwin. I have been in touch with at least six second-minded people as a result of my book, Dark Energy. Only one of them can be objective regarding his "gift".

Ancestry

As a matter of interest, Charles Darwin had three close friends who were essential to the publication of *The Origin of the Species*, one of whom was Sir Charles Lyell, the geologist. I possess very strong circumstantial evidence in favour of my being related to Sir Charles via his eldest sister Frances, very probably my third great grandmother. Frances had a boy child, James, with a Sir James Carnegie who was at the time of 1816 the Baronet of Southesk. That child became my third great grand father, James Carnegie, on my mother's side. My mother's maiden name was Vera Louise Carnegie. The burial sites of both James Carnegie and Frances Lyell have been positively identified and DNA tests could be carried out on the bones of the presumed mother and son. If positive, this might explain how I've been able to come up with such strange ideas regarding the so-called universe!

The Salvation Of Jesus

As unique individuals, we are capable of living in complete harmony with each other and with Reality and the spirit of Reality. Then we will then all realise our hearts' desire. The universe that we perceive as real in the form a vast cosmos is full of fascinating objects such as stars, galaxies, black holes, planetary systems, maybe even "white holes" and the vastness of space and time on the macroscopic scale, as well as an array of atoms, photons and subatomic particles and their associated fields on the microscopic scale. Although real they have all been "engendered" into being by the subliminal imagination of humanity itself.

Unconscious imaginations, both individual and collective, are continually searching for new "ideas". When they come up with one that lowers the level of conflict within the UB it is automatically installed and later "discovered" by conscious minds as a new potential "fact". Science is at least sincerely trying to discover the "truth" about the world in which we live, in the greater sense the universe. But it is also "creating" the facts that it later "discovers" as additional features of reality in the form of all those PhDs running around out measuring and analysing everything they can lay their hands on. I have respect for them all and I study their work but if they don't realise what they're really looking at, a universal belief as opposed to an objective universe, it will all be in vain because it amounts to a delusion. Studying the universe is not a democratic process. The so-called scientific consensus is not what it's made out to be. We need to wake up and get with Reality or we will all suffer the "logical" consequences, as did all the civilisations within the UB before us. The only way to deal with it is to allow our polarised mind, the conscious and unconscious, to become one fully unified integrated mind. Such a mind will inevitably operate below the level of awareness, giving us the sensation of living in the so-called Holy Spirit: The spirit of Reality. This is actually the salvation that Jesus of Nazareth is trying to get across which currently is being derailed by the various religions driven by the so-called second-minded phenomenon. Jesus wept!

This state of consciousness beyond all religion makes us all, Jews, Christians and Moslems, children of Abraham, if we must use theological terms. I daresay that many people experience much peace, comfort, joy and strength by having "faith" in any one of the great religions, not just the ones mentioned above, but it's nothing compared with the Real thing; living in Reality as unique individuals in perfect harmony.

Second-Mindedness in The Clergy

As mentioned, one in twenty humans have a second mind within the other half of their brain. More than one in twenty clergymen possess this trait, unsurprisingly. It makes them feel that they have a closer contact with the spirit of Reality than the rest of us, although it is virtually impossible for anyone with this family trait to be objective about it. There are, however, "side effects". Where clergy cannot marry, sexual perversions may occur because celibacy is unnatural and causes stress. The spiritual host will get concerned about its physical host's state of health and well-being and the effect that it may have on them. We need to realise that although the second mind is living very much closer to Reality, it can't actually live *in* Reality because it shares certain functions with its physical host. On occasions, the second-minded spiritual host will be concerned about its own state of being, even its very survival, and can therefore suggest activities that are generally considered immoral, such as sexual abuse and even child abuse. I'm afraid that nowadays there are too many like those Pharisees of the Bible; all rules and regulations and very little love.

The only solution to such a problem is for the gifted person to get closer to Reality by paying heed to the words of Jesus, in the Christian tradition, and renouncing everything, in particular his conscious mind with its knowledge. This allows it to merge with the unconscious mind and become one integrated "logical" system; the salvation of Jesus. This will allow the second mind to literally attain Reality so that it is no longer concerned about its own survival, solving the problem for all concerned. This will enable damaged persons together with the damaged community to be healed by what some will consider supernatural healing powers emanating from the newly upgraded clergyman, to everybody's complete satisfaction. Otherwise, it will become a terrible social dilemma with no adequate solution, just bitterness, hatred and perplexity leading to further deterioration in society. This is the lesson demonstrated by Jesus when an angry group of Pharisees brought him a woman who had been caught in adultery. This was a very serious offence in those days, punishable by being stoned to death. We laugh at such things today because such behaviour now seems normal and alright. Today, when deviance concerns child abuse, the same public abhorrence results, because the people have no faith that the situation can be healed. We need to ask what would happen if a person who had abused children was brought to Jesus or someone like him. When we think about this issue in the light of the theory of the universal belief we can understand why a similar scenario would occur, whereby both the children and the community involved can be completely "healed" by such a charismatic figure. The healing scenario would be installed into the universal belief. It would be as if the abuse had never happened, and all those involved would experience no ill effects at all. They would either have no memory of the event ever taking place or they would realise the loving, healing power of the spirit of Reality and be completely amazed. This is quite different than the conventional way of dealing with such problems, whereby hard-hearted people get involved, permanently damaging the entire civilisation that, in conjunction with many other unRealistic activities and thoughts, becomes doomed to annihilation like every civilisation that has gone before. Jesus or anyone else living very close to Reality would weep at such behaviour.

The Direct Interaction with Reality

However, those of us who are not gifted with a second mind can realise a direct interaction with Reality by identifying our being with the ERS instead of the real self. This can only be achieved by allowing our conscious mind to upgrade and merge with the unconscious mind, creating one integrated "logical" mind that operates below the level of awareness, producing a sensation of living in the spirit of Reality as an integrated part of Reality. In fact, those so-called gifted individuals also need to achieve the same integrated state of mind, which will allow not only themselves but also the second-minded intermediary to upgrade even closer to Reality. In this situation, the intermediary will not be threatened by such an upgrade and so will not react negatively against those who suggest such a thing. Instead, it will encourage its host to participate in such a course of action.

So-called supernatural events emanating from individuals can only happen if the level of conflict within the UB is reduced by the enhancement, which it usually is. Such enhancement is usually perceived as a social healing event. But because the second mind shares certain functions with its host, it is not actually living close enough to Reality to be immune from concern regarding its own survival when the chips are down. This means that such interactions sometimes become contaminated by speculation from the host mind, resulting in conflicting religions. It is not until the host individual allows his real self, together with "his" ERS, to upgrade very much closer to Reality, as mentioned above, that both the host and second mind are freed from the coercive interaction and can be "trusted" by Reality. Unless the upgrade is allowed to occur, the second-minded family trait inadvertently fashions how many of us are led to think about God as an almighty "out there somewhere". If we can see beyond the second-minded phenomenon, we ultimately realise that the one true God is the spirit of Reality for all mankind, regardless of which religion is currently embraced and whether or not we have a second mind. Any individual real self, together with its ERS, is able to upgrade very much closer to Reality by allowing its polarised mind, the conscious and the unconscious, to become one fully integrated "logical" mind that inevitably functions below the level of awareness. This is actually the so-called salvation of Jesus and is beyond even Christianity or Christian "faith". Such an individual still has a very simple but adequate child-like conscious mind that is in no way childish. The driving force behind such an individual comes from the fully integrated mind below the level of awareness. This is why such a state is considered to be living in the spirit; the spirit of Reality.

Hypnotic Susceptibility and Second-mindedness

In passing, we need to be aware of another family trait that goes hand-in-hand with second-mindedness, and also occurs in about one in twenty, making an individual mind hypo-sensitive. This is so-called hypnotic suggestion. Such individuals are referred to as "deep trance hypnotic subjects" and this group is responsible for so-called UFO sightings and extra-terrestrial encounters and, in conjunction with second-mindedness, angels. Such individuals seem to be able to spontaneously hallucinate anything that their imagination is infatuated with. It is actually the second mind that acts as a so-called hypnotist, hypnotising the host mind into a delusive perceptive state because the second mind "knows" that this is what the host mind wants to think and believe. This is also the realm of schizophrenia. If those who claim to have seen UFOs and extra-terrestrial life-forms are tested for hypnotic susceptibility it is discovered that they are all so-called deep trance hypnotic subjects unless second-mindedness is involved. As mentioned above, hypnotic hallucinations can become real for a group of "normal" people without these gifts if a "charismatic" second-minded person is involved in the observation. A mass hallucination will result, strongly reinforcing for many people that UFOs and extra-terrestrial life-forms are real. However, such events will not be authentically real until all of humanity is involved in what will be known as a first contact event. Furthermore, observation will be lacking in technical details at this stage of UFO development within the unconscious imaginations of those at the forefront and sightings will be rather vague in their technical makeup.

The Psychodynamics of Second-mindedness

We need to a second look at the so-called second-minded phenomenon. Because it can be so easily misunderstood, we'll go through it all again in a little more detail. Second- or double-mindedness arises within a human brain because it is more or less in two halves, the right and the left. The two halves are connected together by a bundle of nerve fibres referred to as the "corpus callosum" which contains relatively fewer nerve fibres compared with the neural network that makes up the two brain hemispheres. This means that, in some areas of each half brain, there is a slight degree of autonomy. The personality centre of an individual human mind must be in one side of the brain or the other, but not in both. Sometimes, however, a separate centre point starts to develop in each half of the embryonic brain. Usually one centre point will become redundant and just "fizzle out", leaving just one mind to develop within the whole brain. On rare occasions, both centres develop side-by-side, with one playing a dominant role. This becomes the mind of the person involved, the dominant or physical host mind. The other centre develops a totally separate second mind below the level of awareness of the dominant mind, the so-called physical host. This is above and beyond the "normal" separate working of the two halves. Effectively, it amounts to two personalities in the same brain. The phenomenon of double or second-mindedness is a family trait passed down from generation to generation. The second mind is living closer to Reality than its physical host, due to its simplicity, although it is not living in Reality because it shares many functions with its physical host.

This means that the second mind is still concerned about survival, although not as strongly as its host. Years can pass without the second mind making itself known to its physical host, although there is a low level of conflict between the two minds that steadily increases, causing the physical host mind to become more unstable over time. At the highest point of instability, the second mind can break through and make its awesome presence known to its physical host as a divine figure of some kind, due to its proximity to Reality. At this point, this "divine" God-like figure becomes the spiritual host and the original physical host person now becomes its willing servant or disciple. This is a joyful, uplifting event for the "gifted" person who perceives it as outside of "his" own body or real self. Furthermore, the local coercive interaction between the two minds that cohabit the same brain, like a master and his servant, can enhance the universal belief closer to Reality. This is the essence of scriptural charisma and the realm of so-called miracles. The enhancement is below the level of awareness of all those concerned, who witness events that are considered to be supernatural taking place around them, further reinforcing the beliefs of all concerned that the new spiritual host must be a God of some kind. Such events are usually healing in nature because the level of conflict within the universal belief is reduced. This is what belief enhancement means; it's not just changing one belief for another. The beliefs mentioned are unconscious beliefs much like software running in a computer.

A "man" with this kind of double- or second-mindedness possesses "charisma" and is a so-called Prophet in Old Testament terms because he can create the future within the universal belief. Isaiah is a good example. This kind of interaction with Reality can sometimes become contaminated by speculation from the physical host. This is the root cause of all religious controversy and conflict. It is also the basis of most of humanity's thoughts about God, since the second mind is perceived as beyond the real self and sometimes the space and time round it and it's virtually imposable for such a person to be objective about their gift.

Those who inherit this gift from their family line perceive their new-found spiritual host from a deeper perspective than the rest of us because they tend to identify their own being as the "essence of the real self" ERS instead of their "real self". This is what some people call the "soul" and, indeed, it is the only Real thing about our being. They perceive the new-found host, due to its proximity with Reality, as a divine figure living apart from their "real self". Sometimes the new host is perceived as an Angel, sometimes as a God, depending on location and education, or maybe as the Christian god that has no name or even as Jesus, as St Paul perceived his newly-emerged second mind on the road to Damascus. This is the realm of the Gods and they need to be respected because they are super-human life-forms in their own right that are living closer to the spirit of Reality than any human. They too are concerned about their own survival when the chips are down and will dramatically protect themselves, if need be, by enhancing the UB detrimentally against those who cause them concern. This is what is meant by having a "spell" cast on one. The Real God, if we must use the word God, is the spirit of Reality, and living in the spirit of Reality is the ultimate spiritual enlightenment, beyond all religion.

The phenomenon of a second-minded relationship with Reality goes hand-in-hand with another gift which is also a family trait, the gift of a deep trance hypnotic profile; DTH. About one in twenty human individuals has both these gifts together and is hyposensitive to hypnotic suggestion. These two gifts work together firstly because of the trust and reverence that the physical host has for the spiritual host being, hence the physical host can move closer to Reality developing charisma in "his" own right. In this situation, the spiritual host being/mind is not only able to "hypnotise" the physical host's mind into perceiving a modified personal version of reality according to "his" heart's desire but can also, by working with the real self of the physical host, enhance the universal belief.

The enhanced universal belief is perceived in due course as a "supernatural" social healing event of some kind. From the superficial point of view, the hypnotic effect gives rise to spontaneous hallucinations. Such gifted minds will perceive UFOs and extra-terrestrial life-forms as real entities. It is virtually impossible for them to be objective about their altered perceptions of reality. In days gone by, angels and gods were more in fashion than they are now.

The phenomenon of second-mindedness fashions the ideas and perceptions of God for the vast majority of people who are not gifted in this way, but are inspired by the charisma of those gifted individuals who, from time to time, come into their midst. This phenomenon is awesome to all who witness the supernatural events that it engenders. The occurrence of "supernatural" events convinces many people that they must be caused by God, although not what gifted people say regarding their understanding of them.

So far as the Jewish faith is concerned, this is the Old Testament interaction with Reality, amounting to a proxy interaction with the spirit of Reality in the form of Jehovah, an intermediary between the spirit of Reality and mankind. Jehovah must be respected but not necessarily followed. Even the Christian God that has no name must be respected but not necessarily followed. As a consequence, faith means believing in the existence of such a God, a divine invisible being who exists in the "spirit world", who occasionally talks, in "revelations", to certain individuals, referred to as "Prophets". This kind of interaction with Reality is always contaminated by speculation from the minds of their physical hosts and is added to by disciples of the faith as time goes on, eventually becoming top-heavy with rules and regulations that becloud the spirit and simplicity of it.

With the advent of Jesus of Nazareth, a new kind of interaction with Reality was demonstrated, a direct interaction. At this point, double- or second-mindedness was superseded by a loving, caring, healing interaction with the whole of humanity instead of the rather clannish and sometimes apparently vindictive nature of a proxy relationship as demonstrated in the Old Testament. In order to embrace this type of interaction with Reality, one needs to understand and perceive God as the spirit of Reality, as opposed to the intermediary, proxy interaction represented by Jehovah and the like. When we actually believe in God, not just the existence of God, God becomes real to us. God is spirit, specifically the spirit of Reality, and when the spirit of Reality becomes real we live in that spirit. This is what is meant by a direct interaction with Reality. It is important to realise that, when we upgrade into Reality, it not only changes us as a person but also the world in which we live. Our new, upgraded world interacts with the people around us in a loving, healing way. We are then driven directly by Reality which, in theological terms, is the Mind of the God, with God being the spirit of Reality.

This type of interaction with the spirit of Reality demonstrates that we can never have a fully integrated conscious understanding of the truth or the meaning of life, because all that an individual can ultimately embrace is his Real self which is an integral part of Reality, not the whole of it. This is why Jesus of Nazareth, who was living in his Real self, an entity he referred to as the Son of Man — the essence of Jesus — can be considered to be a Son of God, a smaller part of the whole. From the emotional point of view, it is a bit like a father and son relationship.

This is just one way in which the Universal Belief becomes enhanced. We all have a small measure of charisma in one form or another, and events created with what can be referred to "normal" charisma are just as dramatic when they interacts collectively with Reality, although such events are not considered to be supernatural because of the understanding we share regarding them. Natural "disasters" and technological innovation are but two such manifestations, albeit never as dramatic as so-called Biblical charisma.

Science perceives its inventions, "discoveries" and innovations as real because we come to unconsciously believe in their reality which means that it has "unconscious" faith in its conscious understanding of them. The so-called "discoveries" are actually inventions by the collective unconscious imagination of humanity, especially so-called past events, because these are also an integral part of the universal belief. The so-called coercive interaction is none other than the so-called dark energy that fills the universe. In theological terms it can be thought of as the hand of God, which we now understand to be the spirit of Reality.

There are Four Main Ways in which the UB can become Enhanced

1) By the so-called "normal" interaction within the unconscious collective mind of humanity, which is responsible for creating everything perceived as real, including components such as light photons, which are presumed to have been in existence since the "beginning", whatever that means, technological innovations and the engendering of new "discoveries". We are all involved in this type of interaction.

2) By the aid of a proxy interaction, an intermediary, in the form of a second mind within certain human brains, when viewed from the superficial perspective of the "real self". This is a consequence of the human brain being in two hemispheres, each with a degree of autonomy. Proxies form a bridge between an individual human mind and Reality. When viewed from the deeper perspective of the "essence of the real self", the ERS, the proxy is a separate individual living closer to Reality than its physical host individual, with a permanent relationship with "him". One in twenty has various "shades" of this interaction as a family trait and it is considered to be a gift. This is the realm of the Prophets, angels and extra-terrestrial life-forms and is responsible for religious spirituality, but it is an interaction that becomes tainted by speculation from the physical host mind which is not living close enough to Reality. This type of interaction with Reality can produce spectacular supernatural events, but for some it is also the realm of schizophrenia.

3) By means of a direct interaction between an upgraded human mind and Reality; this is the realm of spirituality beyond religion. The conscious and unconscious minds become one mind below the level of awareness, perceived as living in the spirit of Reality. This is the interaction that engenders healing supernatural events on a grand scale, as demonstrated by Jesus of Nazareth.

4) By the fragmented "polarised", individual or collective mind of humanity, the conscious and unconscious. This is the interaction that causes all the trials and tribulations of humanity, and sometimes great emotional healing, because the unconscious collective mind creates the future in the form of the logical consequences of the unRealistically motivated conscious collective mind of humanity. Basically, this is because reality is a conflicting logical structure in its current form. When this fragmentation is "healed" we live in our Real self, in the spirit of Reality. This doesn't change just us, but also the world in which we live.

Dark Energy and the D-Weapon

I recently read on Wikipedia about the proposed new directed energy weapons, E-Weapons, consisting of using high-energy laser-beams, high-energy particle-beams and the like, instead of "throwing" chemical explosives and projectiles at a target to destroy it. It is being proposed that beams of energy be used for the same purpose.

"Dark energy", that mysterious energy that fills the universe, has been equated to the subliminal "coercive interaction" between the minds that make up the total collective mind of humanity. In order to understand this, we need to be able to think of the world in which we live, the universe, as a consequence of belief, a Universal Belief perceived as a dream-like world at the seat of consciousness within each individual human brain, as discussed above and in *Where Angels Fear to Tread*. There are many accounts of "deployment" of the "dark energy" as a D-Weapon, in battles recorded in various scriptures. Here, we'll look at just one, from the Old Testament: the Assyrian King Sennacherib's third and final campaign against the region and in particular Hezekiah, King of Judah in 701 BC in the days of the Prophet Isaiah. He reassured his king Hezekiah, as a message from their God "Jehovah" that he (Sennacherib) would not enter their city (Jerusalem) nor shoot an arrow there, nor come before it with shields. "Jehovah" told Isaiah that he would defend the city to save it, for "his own sake and for his servant David's sake." Sure enough, during the night before the next morning's attack could occur, 185,000 solders of the Assyrian army were slain by the "Angel of the Lord". (2 Kings 19:32 NIV) You may have heard this story many times. Thinking about the universe in conventional terms, it doesn't make any "real" sense, unless one happens to be of the Jewish faith or a sympathiser.

In the main text of *Where Angels Fear to Tread* as well as *Dark Energy* this event is considered in a very different light, because I consider the "universe" to be a consequence of belief. The subliminal "coercive interaction" between all the minds of humanity is the "dark energy" that permeates the dream-like universe. Reality is defined as the potential Real collective mind of humanity, at least, and is composed of the sum total of all the Real selves of each individual person. The "Real self", as opposed to the "real self", is a non-conflicting "logical" system that dwells in Reality as an integral part of Reality.

In *Where Angels Fear to Tread and Dark Energy*, this concept and its ramifications are discussed at length and from many different viewpoints. Jehovah is thought of as an awesome second mind in the brain of Isaiah that enabled its host mind (the mind of Isaiah) to move closer to Reality than it could have done if it was just a single mind. This is possible because the second mind is living closer to Reality than any ordinary human mind, as a result of its simplicity. This enables the second mind to enhance the "intimate coercive interaction" between the two minds, moving the mind of Isaiah closer to Reality as a consequence. This is the essence of Old Testament charisma, resulting from what amounts to a "proxy relationship" with the Spirit of Reality, through the medium of the second mind. This is an interaction that becomes tainted by subliminal speculation from the host mind, making it possible for a "social healing" to occur, because the level of conflict within the Universal Belief was reduced subsequently by the enhanced, strongly charismatic mind of Isaiah. That second mind was not able to live completely in Reality because it shared certain mind functions with its host and, consequently, was concerned about self-preservation, just like any other life-form not living in Reality. In order for the second mind to save itself, it had to save Isaiah. This meant saving Jerusalem. Remember the quote above: "I will save this city for my own sake and for my servant David's sake." The charismatic mind of Isaiah that killed those 185,000 men overnight, by getting them to believe, inadvertently, in their own demise via the Coercive Interaction (CI). Isaiah, however, cannot be held responsible because all the appropriate "mental" activity that took place was below his level of awareness.

Second-mindedness is a family trait. This implies that Isaiah was a descendent of earlier prophets from the time of King David; he would have inherited the charismatic family trait from that line. Second-minded people always considered themselves to be servants of their awesome second mind. Even those who had no second mind, such as both King David and Hezekiah, considered themselves servants of what their prophets told them about their God, "Jehovah." Those Biblical Kings had a healthy respect for their holy men, the prophets, because of the supernatural events that occurred around them. This was due to their prophets' awesome ability to enhance the Universal

Belief, the consequences of which were perceived, in due course, as "miraculous" events, as no one realised that their beliefs had been changed.

We could also say that the collective mind of the Assyrians became more and more polarised because it was thinking and acting in self-interest further and further away from Reality, enabling the more Realistic unconscious part to enhance the Universal Belief, making it possible for everyone within the UB to perceive the logical consequences of the Assyrians' actions. In this case, it was not just the mind of Isaiah that killed 185,000 men of the Assyrian army, but a component of both types of interaction. The "charisma" of their own polarised collective mind caused the event to occur, in conjunction with the "charismatic" mind of Isaiah, which enhanced the Universal Belief, bringing about the "logical" consequences of unRealistic actions, perceived in due course.

The "dark energy" channelled via the CI could be "used" as a Dark Energy Weapon, a "D-Weapon," as opposed to the 21st century proposed direct-energy weapons "E-Weapons" (high energy lasers; particle beams) and the like. "Dark Energy" in the form of a "D-Weapon" cannot be deployed as can the proposed "E-Weapons." A very much closer relationship with the Spirit of Reality is required before it is possible at all. The D-Weapon can only be used for social healing much as Jesus of Nazareth was able to bring about a social healing when the water was turned into wine at that documented feast at Cana or the faster-than-light transit across the Sea of Galilee from near Tiberius to Capernaum, as mentioned above.

The charisma of Jesus was due to a direct interaction with the Spirit of Reality, like certain minds in science only very much more so. Without a second mind, his mind was living very close to Reality. This is what makes the difference between the old and new covenant relationship with "God." The Spirit of Reality is considered to be God, the coercive interaction the Hand of God, while Reality is the Mind of God, in theological terms.

The D-Energy can only be used in social healing interactions, even if some people do get hurt, or even killed, in a proxy relationship with the Spirit of Reality. When D-Energy is deployed in a direct relationship with the Spirit of Reality, no one is hurt because there is no contamination from a host mind's subliminal residual vindictiveness, as there is in a proxy relationship. D-Energy deployment is perceived in this case as a loving, healing interaction, curing the minds of belligerent adversaries rather like the physical healing experienced by those around Jesus of Nazareth.

The coercive interaction can only enhance the Universal Belief into a reduced noise level state, a reduced level of conflict within its structure, as opposed to increasing it. This means that Dark Energy cannot be used in conscious interactions like the "ordinary" energy deployed in the proposed E-Weapons. Only minds living closer to Reality than those of the rest of humanity can engender such interactions. This is the essence of charisma. Jesus of Nazareth was gifted with charisma on a dramatic scale because he was conceived and born very close to Reality.

D-Energy weapons are inadvertently deployed in modern wars. The side aiming to live closer to Reality always prevails, because the collective minds involved, acting in self-interest, become more and more polarised, making it possible for the unconscious part, either individual or collective, to enhance the Universal Belief via the coercive interaction. This means that the logical consequences of unRealistic motivation are perceived in due course by the whole of humanity within the dream-like world of the UB: The universe. In such cases, they are always perceived as adverse. In wars documented in scriptures such as the Bible, both charismatic and polarisation interactions are involved but D-Energy deployment always enhances the Universal Belief, which becomes realised as the logical consequences of unRealistic motivation.

The coercive interaction is invincible because it creates future events within the Universal Belief, a future that comes to be realised; a future that humankind can have little or no conscious awareness of or control over. This is why we say it is the hand of God, in theological terms. Some people like to think that God is in control, but it would be more accurate to say that Reality is in control. Of course, it is not always easy to ascertain which side is aiming to live closer to Reality because we can only see part of the picture, whereas Reality is the whole picture. *Dark Energy* supplies a more integrated, scientific way of thinking about such events, without theological speculation.

In practice, both theological and scientific understanding amount to the same thing, although scientific understanding clearly shows that it is not possible for a single human mind to contain Reality but only an integral part of it as the Real self. This means that we can never understand the meaning or purpose of what we call life, but we can live in that meaning, which does away with the need for understanding the meaning and the purpose of life or even the need to belong somewhere. Under such circumstances, we have no conscious mind to think of such things and yearn after them. This state of affairs is brought about by allowing the conscious mind to unite with the unconscious mind as part of an integrated logical structure referred to as Reality. This is where we each live as our Real self. This is an unimaginable state of being, referred to in Zen Buddhist circles as "no-mindedness" or a state of having no conscious mind and hence no conscious thoughts. In Christian circles, it is where the Son of Man has nowhere to lay his head. Both the Son of Man and Buddha are templates for the Real self; an integral part of Reality, an unimaginable state of being.

God as the Spirit of Reality

Dark Energy brings together subjects as diverse as physics, hypnotism and religion, or what religion amounts to. It begins by considering the world/universe in the conventional sense, as a matter of "objective fact", meaning that it came into existence sometime in the past and has nothing to do with humanity. This is something we tend to take for granted at a very basic level, never seriously thinking that it could be otherwise. Material objects are real in such a world, because they are components of the objective universe.

The "universal belief" is introduced as a universe is based on unconscious collective belief instead of objective fact. In this case, "material objects" are perceived as real because of a subliminal coercive interaction between all the minds of humanity in what could be called a "perceptual consensus".

Scientists, who are traditionally and inadvertently gripped by the concept of the objective nature of our world, struggle when they try to discover how the universe came about, by thinking and talking about a "big bang" supposed to have taken place some fourteen billion years ago, and how mankind evolved from the chance combination of certain organic molecules within it. On the other hand, when these same questions are asked about a world created by unconscious belief, the universal belief, the answers are interestingly different.

Some may wonder whether God is responsible for everything, but what does this mean? Again, we can take things for granted; it's too easy to assume that we are all referring to the same entity when God is mentioned. If we ask the people around us if they believe in God, some will say "yes" and others "no", but what are they thinking about? Many must be thinking about a more developed version of what they have been "taught" about such things from their childhood. It's not surprising that many say they don't believe that there is a God.

If those who say they believe in God are asked if God is real, they will all say "yes". Before we can really come to grips with such fundamental questions, we need to ask what being "real" means. Are we thinking about the reality of our world or the reality of God? The reality of our world, or simply reality, is a matter of experience for us all. But is it real because we have all come to unconsciously believe in its reality or is it objectively real? One way or another it is "real" for all of us. Whether it's exactly the same reality for each of us is another question. The reality of God is yet another matter. Can God be real in terms of objective fact or belief? God is certainly not real to everybody, as is reality. Many of those who say they believe in God, and say that God is real, are really saying that they believe in the existence of God, a totally different matter. We don't talk about the reality of the universe in this way; it is real for all of us, however we think about it.

There are certain gifted people with a special psychodynamic makeup to whom a manifestation of God is real in the form of a second mind. About one in twenty humans have this family trait. As far as the spirit of Reality is concerned, when we do actually and unconsciously believe the spirit becomes Real and we live in that spirit; the spirit of Reality.

The two main concepts in this work are the universal belief, as opposed to a universe of objective fact, and the coercive interaction between all the minds of humanity, which creates and continually upgrades the universal belief that is perceived as reality. Using these ideas, we develop an understanding of reality as a conflicting logical system in its current form, and show how it can become a non-conflicting logical system defined as Reality. We can also explore how Reality relates to the theological concepts of the various religions.

If we must use the word "God", let it mean the spirit of Reality. Then we will have a Real purpose in life, with a Realistic understanding and perhaps, in due course, get to live in Reality. Actually living in or very close to Reality is the interaction demonstrated by Jesus of Nazareth. These two concepts, in combination with the concept of the VLA, which is discussed below, explain everything that mankind has ever experienced, both natural and supernatural.

In order to keep our feet firmly on the ground, we start at a very basic level by considering the nature of the "material objects" that comprise our world. These are objects that are made of what is referred to as "matter", that exhibit "mass" and "inertia" and feel solid to the sense of touch. The gravitational, electrostatic and magnetic fields that we experience are considered by using the universal belief instead of the objective universe to show that weight, for example, is a consequence of our bodies (real selves) being placed onto the surface of the earth within the universal belief according to the law of gravity. One could say that they were "pushed" on the surface of the earth, within the UB, rather than attracted by some mysterious action at a distance. An unconscious belief is like software running in a computer and provides a very different understanding of gravity to the current one. The so-called dark energy fills the universal belief: the universe is actually the coercive interaction between minds.

Such an understanding is a spin-off from a much deeper understanding of the nature of light beyond the photon and even space and time itself, and the very nature of our individual material bodily existence as a complex material object, the real self being a component of the universal belief.

If everything we perceive as real is a consequence of a perceptual consensus between all the minds of humanity, what about the universe of the past? Surely this cannot also be a consequence of perceptual consensus? We need to realise that the current reality in which we live is just one of a very large array of related realities, referred to as the VLA. These other realities are not elsewhere in space and time or in other dimensions, they are all in the here and now, as there is only here and now in reality as well as in Reality. The difference between the realities that make up the VLA lies in their proximity to Reality, with each subtending a narrow bandwidth of reality. Effectively, we are a community of souls that have dropped out of Reality, and are now drifting around lost in space and time, trying to get back into Reality.

There is an continual drift by migration (dark flow) of the "essence of each real self" (ERF), which some people vaguely refer to as the "soul", between the different realities of the VLA, either closer to or further away from Reality, depending upon the ERS's status in Reality. Within our current reality, the result of such migration is perceived as the death of the real self. The ERF migrates across the reality boundaries of the VLA. On the arrival of an ERF into another reality, it participates in the "resident" coercive interaction within that reality, being integrated into that reality by acquiring a new real self together with its "past". This is how the "past" of the current reality, our universe, came to be. Although this is a very integrated understanding of life, the universe and everything, it still leaves us asking questions if we have the slightest whisper of a conscious mind.

In chapter three of *Where Angels Fear to Tread*, we discuss an experiment in hypnosis that demonstrates how a "new member", a "new real self", is integrated into the current reality, although belief enhancement by suggestion (hypnosis) is no substitute for belief enhancement by the coercive interaction between minds.

One may ask how the theory being expounded in this work can be proven. In fact, it has already been proven if we can consider all the supernatural events in the Jewish/Christian Bible and other scriptures to be authentic. The trouble is that current scientific speculation, which has become fashionable, can't explain such events, so it is assumed that they could not have happened as did, for example, the Battle of Hastings in 1066 or the attack on the World Trade Canter in 2001. This is because science is too tightly gripped by the idea of an objective universe. We don't have to believe anything that we don't understand.

By evolving an integrated theory, even if it is the product of human imagination, that can explain the nature of the gravitational field and also all the so-called supernatural events recorded in scriptures, we gain a more feasible understanding of all the events that have been experienced throughout the history of humanity within the universal belief. More importantly, we see that the human conscious mind can never contain the whole truth that is Reality, because the most that we can have is the Real self, which is only an integral part of Reality. We need the whole of Reality and we just can't have it. We can, however, live in that truth by living in our Real self. This occurs when we allow our mind, conscious and unconscious, to combine, becoming an integrated logical structure below awareness. This is an unimaginable state of being, referred to by some as "living in the spirit" Indeed, it is living in the spirit of Reality.

God from the Simplistic Point of View

What is the meaning of God? This is a question that many people do not consider valid. Those who say they believe in God feel that they do not need to answer such a question and those who don't believe in God consider the question to be totally irrelevant. On another level, some who say that they believe in God do not actually believe in God but rather in the *existence* of God, which is not at all the same thing. Believing that water exists in a desert is one thing. Drinking that water is another. When we do actually believe in God, God becomes real for us and we become changed.

In Dark Energy we ask "What is the meaning of God?" The question is not semantic, but rather personal, global, and life-changing. God is spirit, the spirit of Reality, and when we really believe in God, we live in that Spirit. We need to understand the precise meaning of belief and faith. We tend to use the word "belief" very loosely. For example, we believe it will rain or believe everything will work out. And, when it comes to believing in God, the meaning is very diluted. Why is it important? Truly believing in God results in changing us and changing the world around us. Two people who discuss whether what they believe is good or evil can approach this from their own subjective positions. For one, it may be "good to lie" and for another it may be "evil to lie". What determines the reality behind the question? Neither subjective response is adequate in and of itself, because both "believe" in their positions. However, when you introduce a much larger, objective position, both people are healed when they truly believe. If we really wish to attain our hearts' desires then we need to change our thinking and motivation by allowing our minds and our beings to heal. This change goes beyond religion and science, and it has to start with a choice.

The Meaning Of Life Beyond the Mundane

What is the meaning of life? Why are we here? What is our purpose in life? Many ask these questions, and it seems that few are able to find the answers. In a world that is ever-increasing in complexity, a world of science and technology, a world that has split the atom and explored space, it is unsettling that we still struggle with questions that probe the depths of our souls and beings. In *Where Angels Fear To Tread*, I ask these soul-searching questions and suggest a new way of looking at life, its meaning and purpose. I examine the nature of reality, the meaning of God, and the healing that comes by living in, or very close to, Reality. Reality as opposed to reality is a new concept that needs to be understood in order to attain our heart's desire as a unique individual in harmony with other unique individuals and with Reality.

The book is not about religion, but talks about spirituality in a scientific way. It challenges many of the traditional ways of thought and even the belief systems of our time. It goes beyond religion and even science to help us be true to what we really need to believe in. The answer to the question revolves around the "living fact" that in the here and now I, like the reader, am bestowed with consciousness and will as a "fact" of experience. We all need to ask how we came to be in the so-called here and now. This is the only question that need be answered, and the answer is a so-called living answer and not just an intellectual one.

To "truly" understand the answer to such a question we need to realise that the most integrated "knowledge" that we can ever have is the "logic" contained within the Real self which is only an integral part of truth that is Reality and not the whole of it. Furthermore, the Real self is totally below the level of awareness which means that we can never have an integrated conscious understanding of the truth, meaning and purpose of our lives but we can live in the truth by living in our Real self as an integral part of Reality.

Eternal life doesn't mean living for ever somewhere; it means living outside space and time in the here and now and ultimately in Reality. This is exactly what Jesus of Nazareth is trying to get across once we can disentangle the "message" from all the religious propaganda into which we have become inadvertently indoctrinated by the so-called "education" system under the pretext of faith. After reading this book the reader will come to realise that life *does* have a meaning and a purpose beyond the mundane. Although we are unable to consciously understand that meaning and purpose, we are able to live in that meaning by living in our Real self. In order to live in the Real self we need to allow our polarised mind, the conscious and unconscious, to merge into one integrated mind below the level of awareness.

The Very Large Array of Dream-like Worlds: The VLA verses the Multiverse

Every ERS that has ever lived in this reality is still living in one of the other so-called relative realities, or maybe even in Reality, depending on the status of the particular ERS involved. This implies that all of those realities are expanding too. The entire configuration of realities is referred to as the very large array (VLA) of relative realities, some closer to Reality and others further away, in theological terms more "heavenly" or more "hellish". The status of a particular ERS determines which reality it is living in currently and which it will migrate into when it departs from its current real self. The ERS never "dies" as does the "engendered" real self according to those left behind who consider that the death of a real self has occurred. When it can no longer be supported by the real self it simply migrates from the current reality to another, and then another and another etc., until it attains Reality. We could say that, initially, the ERS are lost in space and time and will remain so until they awake from the dream-like world we call reality and arrive in their rightful place in Reality.

Our vast universe, as if it wasn't big enough already, is only a very small part of a much larger array of other universes, or other dimensions if we consider the universe to be objective fact. The so-called VLA is also a consequence of the collective unconscious belief of all those living in the VLA. It is real not because it is objective fact, but as a fact of unconscious belief because we have come to unconsciously believe in it as a potential extension to the current UB.

The "next" reality, together with the new real self into which the migrating ERS migrates, is engendered into its "physical" form by the "resident" coercive interaction as a different real self than in the previous reality, but with the same ERS. It becomes integrated into the new reality together with a memory of its own past within the new reality in conjunction with other real selves with which it associates as a family member or some other connection. It will be as if it were born into the new reality, and will have no conscious memory of its previous real self because it has no awareness of the nature of its ERS.

By thinking about and believing in the world in which we live to be objective fact, we can be caused to be without meaning or purpose, although the scientists who devote their lives to studying one or more of the various aspects of their universe in very great depth may feel that it does give their lives meaning and purpose. I wonder how deep such feelings are, especially since this kind of approach does not supply satisfactory answers to any of the ultimate personal or collective questions. Are scientists, too, lost in space and time while remaining hopeful?

The Multiverse, Dark Matter and Dark Flow

The so-called Multiverse is a speculation based on the assumption that the world in which we live, in the greater sense the universe, is a consequence of objective fact and represents a multidimensional cosmos. The so-called dark energy that fills the entire Multiverse – VLA – is the so-called coercive interaction between all the minds and ERS's of humanity that "engender" the universe into being. The so-called dark matter could well be the "substance" of the ERS themselves; it isn't actually matter but it is something. The so-called dark flow could well be the "flow" or, as I refer to it, the migration of the ERS that are on the move between the various realities that make up the VLA. In theological terms, the ERS are the so-called souls of everyone who ever lived, not only in "our" current reality but all of the realities that make up the VLA. Some of them are the readers' ancestors and others are mine.

I'm aware that this kind of universe is mind-bogglingly different than the way that we have been "educated" into thinking about the universe. The trouble is that thinking in the usual way omits so-called supernatural events. They are considered to be fictional mythological events and are therefore excluded from the scientific paradigm. Even many scientists lead double lives by being scientists from Monday to Saturday and Christians or some other denomination on Sunday or other so-called holy days.

What this work does is embrace one understanding to explain both science and religion within one single edifice. Otherwise we cannot explain how the universe came to be, or what it is made of and how life is bestowed on inanimate matter before any kind of evolution can occur anywhere on the universe or cosmos, let alone on earth.

During my second year at university, I had an interview with Dr. Steven Black at his office at the Hammersmith Hospital in London. Dr. Black was a microbiologist and, at that time, he was studying skin allergies and the issue of whether they could be treated with hypnosis. He was always on the lookout for the best hypnotic subjects, whom he referred to as "deep trance hypnotic subjects". When I arrived, he greeted me and then introduced me to a young woman who had agreed to take part as his hypnotic subject in a demonstration, which he had arranged especially for my benefit. After a preliminary chat between the three of us which took about twenty minutes seated around a table, he began the demonstration. His subject had been pre-trained to enter the hypnotic trance state when he voiced a single word, which could be anything. No change seemed to have taken place in his subject, who was still sitting at the table, apparently wide awake. He then told her that another man had accompanied me into his office and that he was sitting in the chair beside me, talking with me. He asked her to tell him what we were talking about, and she did. The content of what she said is of little significance, but what is *very* significant is that this subject was hallucinating that fourth person, a man whom she perceived as real, in the room together with the rest of us. His subject did not see him just suddenly appear sitting the chair, no! She not only hallucinated his presence sitting in the chair but also, and this was the whole point of the demonstration, her memory of his entry into the room, together with me. This means that her mind had invented the past twenty minutes of "his" existence. She could have gone over and touched him in order to reinforce the realness of her perception of him, but this would never have occurred to her because she believed totally in the reality of the suggestion given to her by Dr. Black. In fact, she cannot have had any memory of the suggestion being given to her. Everything would have been normal as far as she was concerned. She would have believed that the fourth individual had always been in her world somewhere, this being the first time that she had come into contact with him.

Normally, there is little or no coercive interaction involved in perceptions resulting from hypnotically changed belief states, so they are never considered to be real events. As hypnotised individuals, we hallucinate individually, whereas, within the coercive interaction, we "hallucinate" or "perceive" our world as a group.

Just think: each one of us could have dropped out of Reality a second ago. Our entire world, together with its past, including our individual past, could have been instantly integrated into the resident Universal Belief and perceived as completely normal and real, so much so, that we take everything for granted and never seriously think that it could be otherwise, although everyone around us seems to have a sneaking suspicion that there is something wrong with it!

Disasters and Apocalyptic Events

There is a very sinister aspect to the coercive interaction or the dark energy, whichever way we choose to think about it. Individuals whose minds live closer to Reality than the other minds around them, as mentioned above, will normally, automatically enhance the UB into lower conflicting states which are perceived in due course as creative events or discoveries that are all in effect "healing supernatural" events.

A large collective mind can, however, "bend" the coercive interaction away from Reality for a while, always with devastating "logical" consequences. This is because a large collective mind can become unRealistically motivated, actually increasing the level of conflict within the UB, the "logical" consequences of which are perceived as adverse in some way. This happens because the already polarised mind becomes more and more polarised as the conscious minds move further and further away from the unconscious mind from the point of view of Reality. This means that the unconscious mind is closer to Reality than other minds around it and consequently will enhance the universal belief into a less conflicting state. This is perceived as the "logical" consequence of unRealistic motivation and it will be adverse in some way, maybe devastatingly so. This is the interaction that causes all civilisations to eventually destroy themselves, because they have no understanding of the "true" nature of the world in which they are living. A means is always engendered to bring about the destruction, such as climate change, pandemics, asteroid strikes, seismic activity, gamma ray bursts, solar flares, volcanic eruptions, etc. All the minds of humanity are involved in this kind of interaction. This is actually brought about by the coercive interaction "engendering" the appropriate detrimental reality into existence. There are no un-solicited accidents, disasters or even apocalyptic events; they are all "man" made. The spirit of Reality is a loving, healing spirit that is ready to heal each and every one of us, if we allow the healing to occur.

Even science with all its apparent achievements is becoming more and more unRealistically motivated in its quest to understand everything. It is effectively engendering a fantasy dream-like world into existence which will have devastating consequences; maybe the destruction of all humanity. This sounds like the apocalypse of the so-called Book of Revelation. Second minds are also involved in this interaction and "engender" such apocalyptic events as mentioned in the so-called Book of Revelation. Under normal circumstances the "coercive interaction" is responsible for all the sickness and ill-health of humankind because, although it is the interaction that "engenders" reality into existence as a conflicting "logical" structure for the whole of humanity, it is alien to each one of us and hence causes chronic low level stress which sooner or later makes each of us ill in some way.

Psycho-kinesis

The only way for us to avoid such cataclysmic events is to realise our full spiritual potential by getting closer to Reality. One could think of this as our ultimate state of evolution. Please recall the event in the so-called New Testament in which Jesus was asleep in a boat with his disciples and a storm blew up and started to swamp the boat. His disciples woke him and said, "We are perishing". Jesus stood up and rebuked the storm and instantly there was a complete calm. This is so-called psycho-kinesis and as long as the conflict within the UB is re-duced such an event can occur. If we try to evoke such an event with conscious thought it is unlikely to occur because it would be driven by self-interest that would inadvertently increase the level of conflict within the UB.

Asteroid or Comet Strikes

Imagine using so-called psycho-kinesis to deflect a potentially threatening asteroid or comet strike or even making it cease to exist.

This may sound like science fiction but this is exactly what could happen if just one man was living very close to Reality and that one man could be any of us because we all have that potential. It's a question of renouncing what we believe our selves to be currently.

Such a threatening scenario would be brought about in the first instance by unRealistic motivation of a large collective mind or even the whole of humanity which eventually 'engenders' by unconscious interaction the appropriate logical consequences of such unRealistic conscious motivation into existence within the UB. In due course the engendered 'logical' consequences will be perceived as a real event. This is what causes the demise of al civilisations sooner or later but when the whole of humanity to effect will be global which is the significance of the apocalypse mentioned in the so-called book of revelation.

Psycho-kinesis is not a mind to matter interaction but a mind to mind interaction and as long as the level of conflict within the UB is reduced the interaction will be effected. As it was when Jesus of Nazareth calmed the storm over the Sea of Galilee. That storm would have had several times the energy of the atomic blast that destroyed the city of Hiroshima in Japan at the end of WW2.

Volcanoes and Volcanic Ash Clouds

One might wonder how volcanoes can be 'turned' on and off by the subliminal unconscious collective human imagination? To begin with there are volcanoes in reality that have already been engendered within the UB as part of the bigger picture of so-called plate tectonics, and they're standing ready to be tweaked into activity as the 'logical' consequences of unRealistic motivation. This is what is meant by reality being a conflicting 'logical' structure, and we need to spiritually 'evolve' out of it by getting into or closer to Reality.

There are many volcanoes in the UB laying around ready to be engendered into eruptions within the UB under the appropriate circumstances. As aforementioned there are no such things as random or unsolicited disasters or apocalyptic events they are all 'man' made and befit the unRealistic motivation that eventually engendered the 'logical' consequences into existence within the UB.

We need to bear in mind what has been said above about how the past UB was 'engendered' into being little by little by pattern seeking human unconscious imaginations coming up with new ideas that reduce the current level of conflict within the UB. it's an ongoing process that is taking place in the here and new from micro second to micro second gradually upgrading reality closer and closer into Reality.

Even existing events within the 'past' part of the UB can be updated by enhancement which means that so-called historic events can be changed. This is how so-called dark energy the (coercive interaction) works within the UB. Dark energy seems to be out there in the universe as predicted by the cosmological equations but because of the way we currently think about the universe as a consequence of objective fact we cannot recognise its presents.

Being a conflicting 'logical' structure reality is fraught with awesome dangers that seem to threaten our very existence. This prompts us to look for ways to evolve out of it into a state of perfection and harmony. There is every indication that there is a way but that way seems to cost all that we possess, including our conscious thoughts and all that we consider ourselves to be.

It comes as a shock for the UK and northern Europe to be disrupted by such a thing as volcanic ash clouds but if we could see into the unconscious mind of humanity, which we can't, it would all be so 'logical.' Indeed we could prevent it from happening if we had the mind too, by being Realistic in our wants and desires. One such unRealistic motivation is greed, wanting more and more of everything but when everyone is motivated in the same way it is not recognised as such, everything seems to be alright. Reality is not a 'democratic' system as is reality we either get with it or experience the consequences.

Of course, even volcanoes can be calmed by just one man who is living very close to Reality

Climate Change

Thinking about global warming from the point of view of the understanding expressed in *Dark Energy* is quite interesting because it subtends a deeper perspective than 21st century science can achieve currently, by thinking of the world as a universal belief (UB) instead of a universe of objective fact. We all know about global warming and we think of it as the logical consequence of the large amounts of carbon dioxide being put into the atmosphere by burning fossil fuels such as coal, oil and gas etc., but the root cause is that we all want more and more of everything and it is this unRealistic motivation that polarises the human mind even more than normal. The human mind, both individual and collective, consists of two parts, the conscious and the unconscious, and since we have no direct access to the unconscious mind we ignore it and use only the conscious mind in our dealings with the world and with each other. If the motivation of our conscious efforts become more and more unRealistic the "gap" between the two minds becomes greater leaving the unconscious mind relatively closer to Reality than the wayward conscious mind. Any mind, or even part of a mind, which is closer to Reality than the minds around it, will enhance the UB via the coercive interaction creating future events within the UB which are realised in due course as real events. In the case of global warming, realised enhanced beliefs will be adverse. Belief enhancement is driven by Reality in such a way as to reduce the level of conflict within the UB. The level of conflict can be likened to electronic noise within the UB. No moral standards of any kind are applied.

In order to make complete sense of what I'm talking about, we need to think of the UB as the universe, the world in which we live. The so-called dark energy that pervades the UB/universe/reality is none other the coercive interaction between all the individual unconscious minds of humanity. Not until we understand this simple truth about humanity will we be able to do anything constructive about adverse events caused by the polarisation of human minds.

The same applies to global dimming, which is due to dust particles in the atmosphere. What can we do to resolve this problem? There is only one solution, which is to upgrade the unconscious human mind. Paradoxically, in order to do this, we need to get rid of the conscious human mind by allowing it to "merge" with the unconscious mind to form one integrated logical structure which means that our thought processes will operate below the level of awareness, so that in effect we operate "intuitively" by living in our Real self in the spirit of Reality. This, of course, can only be achieved at the individual level because each individual human will has to be involved, it cannot be achieved en masse by any form of religion. We are only true individuals in Reality, where the overlapping part of each Real self is in total harmony with each other Real self. Under these conditions we no longer possess the urge or need to act in self-interest, to say nothing of self-preservation. Compromise becomes redundant and so does the coercive interaction.

When the collective unconscious mind of humanity becomes un-Realistically motivated it will distort our perception of reality more than usual, resulting in adverse consequences being "engendered" into the universal belief that are perceived in due course as catastrophic events of some kind. For example, global warming is the means, unRealistic motivation is the cause.

Faith

Faith in Science

The reason why science and technology are so successful is because current scientific understanding is in accordance with current reality, at least to a great extent. Our perception of the universe,
the world in which are living, is real because we have come to unconsciously believe in its reality. In science and technology that unconscious belief is in accordance with the conscious understanding of
it. We could say that science has 100% faith in its current conscious
belief and faith that makes it real as opposed to consciously believing
that things are real because they are components of an objective universe. Our perception of reality is due to a mind-to-mind interaction
with Reality and the so-called spirit of Reality activates all the minds
involved.

Faith in God: The Spirit of Reality

When it comes to having faith in God – the spirit of Reality – we
cannot have a conscious integrated understanding of Reality – the
Mind of God – because the most that we can ever have is the Real
self which is only an integrated part of Reality and not the whole of
it. When we do get to live in our Real self we become an integral part
of Reality without needing a conscious understanding of anything or
even a conscious mind. This is what I call "no-mindedness" or "a child-
like mind without being childish". Living in the Real self is achieved by
allowing the conscious and the unconscious mind to integrate into
one integrated "logical" mind that inevitably operates below the level

of awareness and it is this mind that is the Real self, as opposed to real self that we are in reality.

Many people say that they believe in God but what this really amounts to is believing in the *existence* of God which is not at all the same thing. When we unconsciously believe in the spirit of Reality that spirit becomes real for us, and when the spirit becomes real we live in that spirit without the need for conscious understanding or conscious faith.

Many people inadvertently allow themselves to settle for less than the above on account of the so-called second-minded phenomenon: a intermediary interaction with the spirit of Reality that is no match for a direct interaction with the spirit of Reality. This is the realm of the various gods and their religions and it is how many people think about God as an divine entity "out there somewhere". They do not re-alise the truth of the matter. To do so, we must listen to the things that Jesus talks about, allowing our mind to heal and becoming one mind living in Reality.

Anyone who has this kind of "gifted" interaction with Reality, even though an intermediary being is involved, has a much higher level of faith then those without it because they have a strong feeling that their God is real. Their God may talk to them or exhibit a presence or maybe work "spiritual healing". All this contributes to their having real faith, although their understanding of it may be a little vague. Science, on the other hand, can have "faith" in its understanding because the understanding and the reality of the situation are identical, giving the impression that we are living in a universe of objective fact that came into existence in the past and for reasons that have nothing to do with humanity.

Waking up from a Dream-Like World

Within a sleeping dream we are usually in the real self or a slightly modified version of it. On waking, our real self is re-integrated into what it was before we fell asleep, and within the same Universal Belief (UB), the essence of the real self's status remains much the same. Also, the "rules" that govern the dynamics of a sleeping dream are less rigidly adhered to than within the UB. For example, there may be no gravitational field so that we can "fly" like superman.

The Reason for Dreaming while we Sleep

Incidentally, the reason why we must dream while sleeping is to maintain the ability to keep in contact with the coercive interaction. If we lose touch with it, we lose our grip on reality. However, within all dream worlds, objects such as the real self, the ground on which the real self stands together with all other objects around the real self, are usually perceived as "real" and "solid" to the human sense of touch and to each other. They possess inertia and mass for the reasons given, within a hypnotically-induced dream or trance state, whether induced spontaneously or at the suggestion of a so-called hypnotist. Things are perceived as "real" by perceptual consensus within a coercive interaction and not from any absolute viewpoint.

On awaking from a dream-like world, such as the current segment of what has been termed the VLA, the reality of our universe, the ERS "migrates" into a different segment of the VLA and a new real self is created by the resident coercive interaction within that different segment, based on the essence of the real selves, the ERS's status in Reality. Some people refer to this essence of the real self as "the soul."

The process of awaking from a dream-like world is what almost all of us perceive as dying, from the point of view of those left behind within the current "segment" of the VLA. We see it going on all around us all the time and we assume that it will happen to us in due course as, of course, it will. But there is more to dying than meets the eye. To understand the full picture, we need to realise that the "essence of the real self" is the "real" you or I, not the real self; the superficial perception we have of our own individual bodies. The real self is a mere invention of the resident coercive interaction, and is little more than a hallucination from the point of view of Reality.

The new segment of the VLA into which we migrate can be either closer to or further away from Reality, depending on the essence of the real self's status in Reality. The Universal Belief is changed, for better or worse, and thoughts of heaven and hell arise. Reality is heaven, in theological terms, and dream-like worlds are various shades of hell.

The other segments of the VLA are not somewhere else in space and time or even within other "dimensions". As we have already seen, there is no objective universe and no objective past. It follows that there is no objective VLA either. It's all a matter of belief in the form of the Universal Belief (UB). We can only live in one UB at a time; the others are merely potential UB's that virtually overlap here and now. There is only here and now and our perception of the here and now is a distorted version of Reality. This is referred to as the Universal Belief, a belief structure created by a range of human subliminal imaginations.

Since we have no recollection of the previous dream-like worlds through which we have passed, and almost no control over the process, the above situations are more or less by default. Such a state of affairs is likened to drifting, lost in space and time. Once we come to understand this we may well ask if there is a way of getting into Reality in one jump and, in effect, bypassing the above default situations.

So-called material objects are perceived as real because we have come to believe in them by subliminal perceptual consensus. This means that they have been "created" by a range of individual subliminal imaginations within the UB. When the spirit of Reality becomes real and not just real but Real indeed we live in the spirit of Reality, which means that the spirit of Reality becomes Real to an individual person. In theological terms, the Holy Spirit becomes Real. This is an unimaginable state of being, but we need to detach our essence of the real self from the illusory real self by renouncing everything we have both, physical and mental, in order to achieve this state of consciousness. As Jesus of Nazareth said, "Unless a man forsakes everything he cannot be my disciple." "Man" implies a unit of mankind that includes women too. This state is thought of and felt as holy because of the joy, peace, and logical integrity, beyond understanding, that it bestows on an individual mind/person. It is more than being healed emotionally, in the fashion of born-again Christians. The whole mind must be healed in order to live in the spirit of Reality.

When this happens to a person, not only is his mind updated or healed, but his world as well, because the UB is upgraded to, or very close to, Reality. The essence of the real self that gets upgraded very close to the Real self, which is an integral part of Reality. This is brought about by willing our consciousness away from its infatuation with conscious thinking, a state that is referred to as "self-remembering" or "dwelling on the essence of the real self". This is a state of consciousness in which we look at ourselves as we look at another person. When we look at another person we have no idea what he is "thinking" about; we just hear him saying things and observe him doing things. When this state is maintained for a sustained period, the conscious mind and the unconscious minds integrate into one logical system that functions below awareness. This is referred to as a state of no-mindedness in so-called Zen Buddhism. It is identical to living in the spirit of Reality; the Holy Spirit in Western terms. From the Christian point of view the process is synonymous with a permanent state of prayer, specifically the so-called Lord's Prayer that Jesus gave his disciples, saying to them "when you pray, pray like this". It is not a matter of saying the words but maintaining the state of consciousness suggested by them.

In Zen Buddhist circles, a state of selfless meditation in which the essence of mind comes to be realised that is identical to the essence of the real self upgraded to the Real self. Some students of Zen achieve this by focusing the mind on a "Koan" which is any paradoxical statement, usually uttered by a Zen Master. The Koan doesn't make any sense to the intellect, which is why it is able to get the attention away from its addiction with conscious thought processes. There are no words that can truly express the state of living in the spirit of Reality, because it is an unimaginable state of being. Beauty in all its forms gives us a foretaste of what living in the spirit of Reality is like.

Both of the above techniques represent, in effect, waking up from the dream-like world completely, and not just migrating into a different segment of the so-called VLA (Very Large Array of dream-like worlds). It is a matter of allowing the conscious mind used for everyday thinking to integrate with the unconscious mind, into a totally integrated logical system that cannot occur while we are so wilfully trying to think about things in our usual way.

The VLA, like the current dream-like world, the universe, is a product of human subliminal imagination. It does, however, give us a more integrated understanding than we currently have. This goes a long way towards understanding what living in Reality means, and it clearly points the way to our hearts' desire.

Both of the above techniques can be done anywhere at any time without joining or leaving any special group, such as a church or an oppressive government regime. If we continue to run around looking for a special group of people with which to associate, we will continue to be lost in space and time. The sooner we get down to the actual meditation, wherever we happen to find ourselves doing, the sooner we will get into Reality and a life of joy, peace, love and security. Atomic bombs and the like are ineffective in Reality. Atoms, photons and subatomic particles do not exist in Reality. They are not Real, even if they are real to those of us living in the Universal Belief that is the 21st century.

Consider that little story in Judo/Christian scripture, about those three characters, Shadrach, Meshach and Abednego, who were thrown into a fiery furnace by King Nebuchadnezzar because they wouldn't bow down to him and who walked around in the fire for a while and then walked out without even being singed. They were able to do this because fire consists of an interaction between electrons around atoms, and photons, which do not exist in or near to Reality. Those men were living very close to Reality, in the grace of Reality or, in their terminology, the Grace of God, known to them as "Jehovah". It is not possible to live in Reality and within the space and time of the UB simultaneously; that is why it is necessary to talk about living in the "grace" of the spirit of Reality.

The phrase "spirit of Reality" and the word "Reality" cannot be contaminated by religious speculation. We cannot expect Moslems to embrace the Christian way of thinking or vice-versa but we can expect both to be able to think about Reality and, in due course, live in it, when each discipline gains an understanding of what it means and how it relates to their current way of thinking and their current status in Reality. This kind of conversion doesn't involve a loss of faith or face for anyone. The Hindu and Zen Buddhist disciplines have a completely different set of speculations with which to express their faiths, speculations that are quite alien to Westerners and the people from the Middle East. In the final analysis, however, they too could embrace the desire to live in Reality without loss of face or faith. Incidentally, those who talk about faith do not actually believe in God but rather in the existence of "their" God. This is very different and is a very weak interaction with the spirit of Reality. Trying to live in Reality or even just thinking about what it means defuses all religious conflicts.

One may well ask whether or not these church-going Christians who live in a certain way will get to Heaven, as they expect. The answer is, ultimately, "yes", but not in the way they think about Heaven and Hell. If they continue to conduct themselves in the same way within each segment of the VLA through which they pass, they will indeed continue to "climb" towards Reality, which is Heaven in theological terms. They will migrate through each segment until they arrive in Reality. Even migrating from the current segment to a segment closer to Reality will be a paradise compared with the previous one. It will not matter if it takes a thousand migrations to achieve Reality, because the way is continually upwards, with each migration upwards being a paradise compared to the previous dream-like segment of the VLA.

There is no reason why everyone shouldn't get to live in Reality eventually. We could ask if, for example, whether Judas Iscariot will go to Heaven or Hell. After what he did in that segment of the VLA of 2000 years ago, he would have gone down but there is hope even for such as he. The bottom line of the Christian faith is living in the spirit, which is identical to living in or very close to Reality, in the spirit of Reality.

It should be possible to achieve this in one migration by giving up everything, physical and mental. This not only means material wealth but also mental wealth such as our thoughts of tomorrow, yesterday and elsewhere in space and time. As the one Jesus of Nazareth said, "Unless a man forsakes everything, he cannot be my disciple." Jesus is not talking about religion or even Christianity but about living in or very close to Reality.

Proving the Theory of the Universal Belief

One may well ask how this theory can be proven. The direct way to prove it, of course, would be for someone to allow "himself" to be upgraded closer to Reality, into a kindred state to that of Jesus of Nazareth. This is very difficult and most unlikely. Many become faithful Christians but do not understand that, in order to upgrade very close to Reality, one must give up the speculation known as Christianity together with all other speculation in accordance with the advice given by Jesus when he said: "Unless a 'man' renounce everything he has, he cannot be my disciple." This means that we need to relinquish our conscious knowledge and indeed our conscious mind in order to allow the conscious and unconscious minds to merge into a fully integrated mind operating below awareness. So far as Christians are concerned, it means giving up the conscious belief referred to as Christianity that we can so easily become rigidly attached to in the name of faith, hence preventing further "growth" towards Reality. A similar scenario happens in all religions, because most of us seem to prefer to stay as we are in space and time rather than become an integral part of Reality. As the poet Milton said, many people think that "it is better to rule in hell then serve in heaven."

We should always bear in mind that reality is a dream-like-world within our own brain, which means that so-called material objects are made of the same stuff as objects within a human imagination. The world in which we live, reality, is not a dream, but it is a dream-like world. The difference is that the dream-like world we call reality is contained with the whole collective mind of humanity, whereas a sleeping or hypnotically-induced dream are within one individual human mind.

The total perceptual consensus of all humanity engenders the world we perceive as reality. To understand how "imaginary objects" can be perceived as solid requires a deeper understanding of the so-called electrostatic field.

When we ask where our world of reality came from, while in a sleeping or hypnotically-induced dream, it is the same question as where did "this" world came from because sleeping dreams are perceived as real and solid while we are in them. When we eventually wake from a sleeping or hypnotically-induced dream, we realise that it was all in our brain. It is the same with the dream-like world we call reality. This implies that we need to wake from it instead of wallowing in it. That is exactly what happens when we allow our conscious mind to upgrade and merge with the unconscious. This is a state that could be described as living in the "grace" of Reality while living here in reality. Reality is outside space and time so it is not possible to live in Reality and reality simultaneously.

Moving Closer to Reality

There is an interesting way to bring this upgrading process about by allowing one of those so-called second-minded individuals to apply their charisma to do it for us. The charisma emanating from such a person can "heal" another mind and bring it into or closer to Reality. This is what happened to Jesus when he was baptised by "John the Baptist" who was a man with a second mind, if we think about this event from the psychodynamic instead of the usual theological point of view.

This is actually the function of a "pastor" but if a pastor does not have the charisma of a second mind, he will not be able to bring about an upgrading conversion into Reality. All that such a pastor will be able to do is baptise a person into a earthly church that is no match for Reality. In or near to Reality, a church consists of two or more individuals living in their Real self very close to Reality, wherever they may be located in the space/time of reality. The person being baptised must be ready to "let go" his conscious mind with its knowledge and beliefs. If we think about the interaction between "John the Baptist" and Jesus from the psychodynamic standpoint, we can understand that Jesus was born ready; all he had to do was learn about the reality of his day, before he became upgraded very close to Reality by John into the "grace" of Reality, so to speak. Of course, John had to be ready and willing too, which he certainly was.

As mentioned above, those few "gifted" people who have a so-called second mind within their own brains perceive it from a deeper perspective, because they tend to identify their being with the ERS instead of the real self. What they perceive is a second and "divine" ERS outside themselves, which they consider to be the God of their particular religion. In a way, this is true because behind that "divine" ERS is the spirit of Reality which is the Real God, if we must use the word "God", regardless of what religious faith we believe in currently.

What makes the second ERS divine is its propinquity with Reality and its awesome ability to enhance the UB in a way that is perceived in due course as a supernatural social healing event. The reason why such an event is considered to be "supernatural" is because no one involved realises that the UB has been enhanced, as it is a subliminal mind-to-mind interaction.

Proving Biblical Events to be Historic Events

The more likely way of proving the theory is to use it to explain all those so-called supernatural events purported to have taken place in the various scriptures such as the Judeo-Christian Bible which upgrades such events into real historic events, proving their reality and hence this theory of the universal belief.

The Speed of Light and the Retina Optic Nerve Combination

There is another possible way to prove the so-called Theory of the Universal Belief and that is to understand the workings of the human eye. As mentioned above, there are 120-million photoreceptors on the surface of the retina and only 1.2-million fibres within the optic nerve to convey the signals from the retina to the consciousness within the brain. This means that the signals from the retina need to be processed by a device behind the retina before they can be transmitted along the optic nerve to the consciousness of the ERS within the brain; a sort of multiplexing action that will work at a certain frequency. The value of that frequency will slow down the direct consciousness perception of the conscious eye of the ERS to the so-called speed of light perceived by the eye of the real self within the UB. It's a matter of discovering that frequency within the eye mechanism and calculating what the speed of light should be according to the theory of the UB being put forward, if the calculated speed of light is identical with the observed speed of light i.e. about, 186,000 miles per second. This could be used as proof of the theory of the UB and that we are not living an objective universe.

What as been said above is only the tip of the iceberg, because it involves just about everything and everybody.

It is indeed a vast subject.

Glossary

Dark Energy: The coercive interaction between all human minds; together, the Mind that is Reality.

Dark Matter: The "substance" of the sum total of all the ERS's within the VLA.

Dark Flow: The movement of the ERS's between the various segments of the VLA.

reality: The world in which we live: The universe.

Reality: The total potential collective unconscious mind of humanity when each individual mind is a non-conflicting mind within itself.

The real self: The "physical" attributes of an individual person in the current reality.

ERS: The essence of the real self.

Real self: The ultimate potential real self, an integral part of Reality.

VLA: The sum total of all the realities similar to the current reality, either closer to or further away from Reality.

Second-mindedness: A second personality in the other hemisphere of certain human brains.

CPSIA information can be obtained at www.ICGtesting.com
Printed in the USA
BVOW09s1031101114

374431BV00035B/1444/P